SCHOOL of Assassins

The Case for Closing
the School of the Americas
and for Fundamentally Changing
U.S. Foreign Policy

Jack Nelson-Pallmeyer

Maryknoll, New York 10545

Third Printing, August 1997

The Catholic Foreign Mission Society of America (Maryknoll) recruits and trains people for overseas missionary service. Through Orbis Books, Maryknoll aims to foster the international dialogue that is essential to mission. The books published, however, reflect the opinions of their authors and are not meant to represent the official position of the society.

Published by Orbis Books, Maryknoll, NY 10545-0308

Manufactured in the United States of America

ISBN 1-57075-134-X

Contents

Acknowledgments

I want to thank Father Roy Bourgeois, Carol Richardson, Vicky Imerman, and others at SOA Watch for providing me with the extensive documentation that is the basis of this book. Thanks also to Peter Thompson, Eleanor Yackel, and others for various kinds of support. Thanks, finally, to my family — Sara, Hannah, Audrey, and Naomi, who always give me the inspiration, time, and grounding I need to write from my heart about things that matter.

Introduction

Pizza, Torture, and the School of the Americas

I cannot eat pizza without thinking about the U.S. Army School of the Americas (SOA) and its cozy relationship with torturers. To explain this admittedly bizarre association I must tell you about my first visit to El Salvador in the summer of 1982. My wife, Sara, and I were participating in a two-week intensive study tour to Mexico and Nicaragua sponsored by the Lutheran Church and organized through the Center for Global Education of Augsburg College.[1] A logistical quirk in scheduling meant our group had a four-hour layover at El Salvador's International Airport. Amazingly, the Center turned these few hours of potentially wasted time into a life-changing experience involving pizza, torture, and (although unknown to me at the time) the School of the Americas.

After clearing customs our group exited the airport quickly and began the forty-five-minute journey to San Salvador in rickety vans under the watchful eyes of the ever-present soldiers. At this time and for many years thereafter, El Salvador was a place of fear, deadly repression, and crucified hope. Human rights violations, including mutilated bodies left on street corners and disappearances at the hands of military and paramilitary forces, were facts of daily life — and death.

Within minutes we found ourselves at the toll booth

where a year and a half earlier four U.S. churchwomen had been stopped on their way home from a pastoral conference in Nicaragua. Maura Clarke, Jean Donovan, Ita Ford, and Dorothy Kazel, who had worked among El Salvador's poor, were taken a mile or so off the main road, brutally raped, and murdered.

It is hard for me to describe my feelings that day or how it is that four hours in El Salvador could have a profound impact on my life. Even as the plane set down I began to be drawn into the pain and the purpose of the struggling people of El Salvador. Leaving the airport, setting foot on Salvadoran soil, stopping at this particular place, and remembering the murder of the churchwomen filled me with conflicting feelings. Outrage, despair, courage, hope and fear battled within me.

Upon our arrival in San Salvador we met with women who had formed an organization, "The Mothers of the Disappeared." United first by their shared grief over the loss of loved ones, these mothers demonstrated a stubborn determination not to let their children die in vain. Their decision to organize came during a meeting with Archbishop Romero who told them that the Salvadoran government's wave of terror against the civilian population would not be halted through personal grief but through collective action. Romero himself was assassinated not long after their meeting while saying Mass in the chapel of a cancer hospital.

Each woman we spoke to that day had a graphic story to tell of a son or daughter taken by "death squads." These squads, made up of men from paramilitary and military groups, were allied with the U.S.-trained security forces. At the hands of the death squads some of these women's children were "disappeared" — that is, taken from sidewalks or buses, schools or homes, never to be seen again. Others were found dead, disfigured, and brutally tortured, their bodies left in garbage dumps or on street corners where they served as a public deterrent.

It was unfortunately lunch-time when we arrived at the office of the Mothers. With no other time to eat and at the insistence of the Mothers who themselves ordered us pizza, we shared food along with their stories. They told us about their sons and daughters and showed us graphic photos of hundreds of dismembered bodies. They shared their lives, their pain, their hope, and our responsibility. We ate pizza. We looked at grisly photos. We listened to testimonies. Many of us wept, hearts open, pizza in hand.

Sharing food and drink while hearing stories of betrayal and broken bodies, we celebrated "communion" and emptied ourselves of the illusions which filled many of us who had lived too long under the influence of the official rhetoric of the "benevolent superpower." Our hearts were ripped open by a spirit calling us to hope and conversion. In a moment so powerfully poignant in its miraculous absurdity, we ate pizza as one of the mothers asked us upon our return home to change U.S. policy while opening her blouse to reveal scar tissue where her breast had been severed by the machete of a Salvadoran soldier.

— ❖ —

Despite the assassination of Archbishop Romero, the rape and murder of the U.S. churchwomen, the death, torture and disappearance of thousands of teachers, church workers, students, union organizers, and campesinos, U.S. support for a series of repressive governments in El Salvador held firm throughout the 1980s. Thousands of declassified State Department, Defense Department, and CIA documents confirm much of what the Mothers told me on my first visit to El Salvador, including the charge of

U.S. involvement and complicity with El Salvador's death squads. According to an article in the *National Catholic Reporter* (*NCR*), the documents show "that the Reagan White House was fully aware of who ran, funded, and protected the El Salvador death squads of the 1980s, and planned the 1980 death of San Salvador Archbishop Oscar Arnulfo Romero."[2]

This report in *NCR* tells only part of a more troubling story. More than a decade passed after my first visit to El Salvador and the death toll mounted to about eighty thousand before I and many others learned that *most of the men who had planned and carried out these murders and atrocities were trained at the U.S. Army School of the Americas (SOA)*. More disturbing still, El Salvador is only one among many cases dating back to the 1940s and continuing up to the present in which SOA graduates have practiced torture, overthrown legitimate governments, carried out massacres, terrorized civilians, and targeted progressive church workers. "In fact," in the words of a *Los Angeles Times* editorial urging that SOA be closed, "it is hard to think of a coup or human rights outrage that has occurred in [Latin America] in the past 40 years in which alumni of the School of the Americas were *not* involved."[3]

School of Assassins is about the U.S. Army School of the Americas: what the school does and why; why it must be closed; and how you can help to close it. It describes the deadly role the SOA has played throughout the Americas over the past five decades. It lays out and then counters the positions of the school's defenders. Equally important, it examines how the priorities of the school and the conduct of its graduates, including an especially targeted terror against the progressive churches, are consistent with and carried out in service to broader U.S. foreign policy objectives that must be challenged and changed.

I ask the reader to keep three things in mind: *First, the School of the Americas is part of a monumental scandal and it must be closed.* Once we clear a path through the mountains of deceptive rhetoric we are left with only two realistic possibilities in assessing the SOA:

> Option 1:
> The fact that numerous SOA graduates are associated with repression, torture, human rights abuses, the overthrow of constitutional governments, targeted persecution of progressive religious, and other atrocities indicates that *the SOA is a colossal failure and has done its job exceedingly poorly. Therefore, it must be shut down.*
>
> Option 2:
> The fact that numerous SOA graduates are associated with repression, torture, human rights abuses, the overthrow of constitutional governments, targeted persecution of progressive religious, and other atrocities is evidence that *the SOA has tragically succeeded in carrying out its mission and done its job in an efficient, albeit deadly manner. Therefore, it must be shut down.*

Whether you choose option 1 (gross incompetency) or option 2 (deadly efficiency), the School of the Americas can, must, and will be closed.

Second, your efforts are important and can make a difference. Closing the School of the Americas will not be easy. The school has powerful supporters who, despite SOA's clear record of abuse, offer sophisticated defenses. I believe that the weight of evidence against the School of the Americas is so compelling that its defenders won't ultimately be able to justify the school's existence or withstand the weight of public disdain. A movement coordinated by SOA

Watch to close the School of the Americas is building.[4] If you add your voice to the chorus of SOA protest then the movement gets stronger. We can win. I really believe this. That is why good and decent people have engaged in civil disobedience at SOA headquarters in Fort Benning; why priests and nuns and other religious are speaking out and protesting; why newspaper editorials throughout the country are calling for a shutdown of the SOA; why I have written this book; and why I'm asking you to help close the School of the Americas.

Finally, closing the SOA is vitally important but it is not enough. The School of the Americas is, in the words of SOA Watch founder, Father Roy Bourgeois, "a school of terrorists . . . that brings shame upon our country and upon us and our laws and what we should be standing for."[5] Whether the SOA is a tragic failure or a despicable success, its closure must be part of a fundamental reassessment and redirection of U.S. foreign policy. Without vigilance and vision the SOA could be closed through our efforts but the practices that have led it to be known as the "School of Assassins," the "School of Dictators," and the "School of Coups" could be carried out elsewhere within the broader U.S. foreign policy establishment. If that happens, then closing the SOA will be a victory more symbolic than real. For the sake of all of us who live in the Americas we must not let that happen.

Chapter 1

Truth and Consequences

Opening Arguments in the Case against the SOA

> The U.S. Army School of the Americas... is a school that has run more dictators than any other school in the history of the world.
>
> —REPRESENTATIVE JOSEPH KENNEDY[6]

Jesus promises that the truth will make us free. Pilate asks, "What is truth?" and orders Jesus crucified because their understanding of truth differs profoundly. Truth, it seems, is a flexible word with multiple definitions and deadly consequences. I find it troubling and illuminating that Jesus' words concerning truth and freedom are engraved on the wall of the CIA's old headquarters. It is ironic, given the fact that the CIA, like the School of the Americas, has artfully used secrecy and deception in carrying out its sometimes deadly mission.

Salvadoran poet Ramón del Campoamor could have had the controversy surrounding the School of the Americas in mind when he wrote that "in this world of betrayal there is nothing true or false. Everything depends on the color of the crystal through which one gazes." Getting at the truth concerning the School of the Americas depends on sifting through competing perspectives, examining crystals of different colors. Critics and supporters see the school, the world, and the truth through different lenses. They would agree, however, on this background statement from an official School of the Americas brochure:

> The Latin American Training Center-Ground Division was established in 1946 in the Panama Canal Zone. It became the U.S. Army School of the Americas (SOA) in July 1963 with Spanish as the official language. In October 1984, the school suspended operations to comply with the terms of the 1977 Panama Canal Treaty. Three months later, SOA reopened its doors at Fort Benning, Georgia, as a member of the U.S. Army Training and Doctrine Command (TRADOC). SOA has trained nearly 59,000 Latin American military, policemen, and civilians.[7]

To this modest base of agreement we can add the common recognition that the school is funded by U.S. taxpayers and that through the School of the Americas the United States has significantly influenced events and institutions throughout the region by fostering close relationships between U.S. and Latin American military leaders.

Perspectives on the School of the Americas diverge from here like a fork in the road. Supporters portray the school as a fountain from which democratic values and respect for human rights flow. Fed by an ever-flowing stream of U.S. values and good-will, the enlightened curriculum of the School of the Americas fills tributaries throughout Latin America with hope for development and democracy. The school's mission, according to the SOA brochure, "is to provide doctrinally sound, relevant military education and training to the nations of Latin America; promote democratic values and respect for human rights; and foster cooperation among multinational military forces."[8] Among the many benefits and strengths of the school, according to official SOA literature, are the following:

- Twenty years ago, most Latin American countries were under some form of repressive rule. Today all but one... are democratic societies.
- SOA graduates serve in all levels of Latin American government and military.
- SOA... [is a forum that] allows for open discussion on current issues for the hemisphere.
- Students are exposed to American democratic values and ethics.[9]

An information paper from the School of the Americas which seeks to silence SOA critics boasts that because of the school and because of "active U.S. military engagement... Latin America is today the least militarized and least violent region in the world."[10] "All courses," the paper argues, "provide a forum in which military and civilian personnel from the entire Latin American region involved in defense-related issues come together to exchange ideas and create common bonds and respect." All instructors at the school "are graduates of the School's Human Rights Instructor Certification Course and are prepared to discuss human rights issues as well as integrate human rights and related training" into other courses. "The School," the paper asserts, strengthens "the nascent democracies in Latin America" through the "Democratic Sustainment Course which is designed to introduce and teach theory and practice of military and civilian leadership in a constitutional nation-state," and through a course on "Peace Operations" which is designed "to train the student in emerging U.S. doctrine for peace operations strategies."[11]

This rosy picture is framed by claims that events throughout the 1980s in El Salvador and Honduras offer profound testimony of SOA success, not failure. Colombia, according to the paper, is another success story, as is Bolivia under the dictatorial rule of General Hugo Banzer. These and other examples of positive results of U.S. policy in Latin America are attributed to the enlightened philosophy and practices of the School of the Americas.[12]

Critics see the School of the Americas in a different light. We will examine SOA curricula in more detail in later chapters. For now it is sufficient to note that it has much more to do with counterinsurgency than democracy, with fighting dirty wars than peacemaking. The "Democratic Sustainment Course" mentioned above, for example, wasn't a part of SOA curricula until 1990, forty-four years after the school first opened its doors in Panama. SOA graduates in El Salvador, Honduras, Colombia, Bolivia, and elsewhere throughout Latin America have stained the soil with blood and have been linked so consistently and perversely to human rights abuses, dictatorial rule, torture, and disappearances that the School of the Americas itself is finally being scrutinized. After decades of operating within the guarded shadows of secrecy, horrific revelations concerning the school's ideology, curricula, and practices are reaching the public for the first time because of the faithful witness of SOA critics. What is revealed when we expose the conduct of the school and its graduates to light is not a pretty picture, and it is in sharp contrast to the portrayal of the school as a bastion of democracy.

In opposition to the distorted picture of truth offered by SOA defenders above, it seems appropriate to begin with El Salvador's Truth Commission. In El Salvador, which means "The Savior," Jesus' words about truth and freedom were taken to heart during negotiations to end a decade-long, bloody civil war that had claimed the lives of approximately eighty thousand people. The United Nations, which played an important role in El Salvador's peace process, was authorized by the warring factions to prepare a "Truth Commission" report as part of the cessation of hostilities. Seeking the truth about El Salvador's dirty war, as we will soon see, led investigators frequently to the doorsteps of the U.S. Army School of the Americas.

The logic behind a Truth Commission was that an honest accounting of who had done what to whom was necessary in order to begin to heal the multiple wounds of war. Honesty, although not as radical in its implications as repentance, was seen as a vital first step in a national healing process. Repentance requires going from truth-telling to empathy, from acknowledgment of deed to remorse. One can't get to repentance, however, without first passing through the treacherous waters of honesty.

It was hoped that by telling the truth about what happened in El Salvador during the course of the war an atmosphere more conducive to forgiveness than vengeance could be fostered. In short, El Salvador's precarious longing for peace had been shaped, even deformed, by a violent, bloody war. However tempting it must have seemed to encourage a national amnesia, those who pushed successfully for a Truth Commission understood that the country could not realistically expect to move into a more hopeful future without an honest assessment of the past.

This final point should be kept in mind because it is true for us as well. At the present moment dishonesty and deceit, secrecy and lies, deception and treachery flow like tainted blood from the diseased heart of U.S. foreign policy, through the hardened arteries of the School of the

Americas, and into the veins of Latin America. For our own sake and for others we need to take an honest look both at the school and at the foreign policy that gives it legitimacy.

On April 30, 1990, almost three years prior to the release of the Truth Commission's findings, a U.S. Congressional Task Force reported that the men responsible for the November 16, 1989, massacre in El Salvador of six Jesuit priests, their housekeeper, and her teenage daughter were trained at the School of the Americas at Fort Benning, Georgia. In August that same year, SOA Watch, the group coordinating efforts to close the School of the Americas, opened an office just outside Fort Benning's main gate. Members of the group did research and carried out fasts and protests in an effort to call attention to the troubling relationship between the School of the Americas, SOA graduates, and torture.[13]

The faithful witness and action of SOA Watch supporters got a boost on March 15, 1993, when the United Nations' Truth Commission released its report on El Salvador. For anyone paying close attention to El Salvador in the 1980s the Truth Commission offered few surprises. As expected, it held the Salvadoran military and a series of U.S.-backed governments responsible for the vast majority of human rights violations, massacres, and civilian deaths. Although the Truth Commission didn't address the role of the United States specifically, many of us felt that our country too needed a large dose of honesty and truth-telling. The United States, after all, provided financial, logistical, military, and ideological support to the groups named by the Truth Commission as most responsible for disappearances, repression, massacres, and other human rights abuses. The report revealed the following:

- Two of three officers cited in the assassination of Archbishop Romero are SOA graduates, including death squad founder and leader Roberto D'Aubuisson.
- Three of five officers cited in the rape and murder of the four U.S. churchwomen are SOA graduates.
- Three of three officers cited in the case of two murdered union leaders are SOA graduates.
- Ten of twelve officers cited as responsible for the massacre of nine hundred civilians at El Mozote, perhaps the single most horrific event of the bloody civil war and a massacre the U.S. government actively covered up, are SOA graduates.
- Two of three officers cited in the case of the El Junquillo Massacre, three of six cited for the Las Hojas Massacre, and six of seven named as responsible for the San Sebastian Massacre are SOA graduates.
- Nineteen of twenty-six officers cited in the November 1989 murder of six Jesuit priests, their housekeeper, and her daughter are SOA graduates.

Overall, more than two-thirds of the more than sixty officers cited for the worst atrocities in El Salvador's brutal war are alumni of the School of the Americas.[14]

But, as it turns out, the disturbing linkage between the U.S. Army School of the Americas, Salvadoran officers trained there, and grisly human rights abuses is just the tip of an SOA iceberg. Long hidden beneath an ocean of secrecy is a consistent and damning pattern of abuse carried out throughout Latin America by SOA graduates since the school's opening in 1946. It is to these issues that we now turn.

Chapter 2

Beyond Coincidence

Toward an Airtight Case to Close the SOA

> [If the SOA] held an alumni association meeting, it would bring together some of the most unsavory thugs in the hemisphere.
>
> —Representative Martin Meehan[15]

Latin American nations with the *worst* human rights records have consistently sent the *most* soldiers to the School of the Americas. Examples include: Bolivia during the reign of terror of General Banzer; Nicaragua during the Somoza family dictatorship; and El Salvador during the period of bloodiest repression. All were top clients of the SOA during the periods in which military abuses were most pronounced.

In my view, SOA Watch has compiled enough detailed information linking the School of the Americas and its graduates to abuses to make an airtight case for closing the school. Presenting *all* the evidence would be nearly impossible. It would take volumes and overwhelm the reader both practically and emotionally. As Representative Martin Meehan of Massachusetts notes above, unless you enjoy conversations with thugs, murderers, and torturers, you'll probably want to pass on any invitation to attend a gathering of SOA alumni.

As we have seen from the case of El Salvador, there

are extensive links between the SOA and Latin America's most notorious human rights abusers. SOA graduates have played key roles in nearly every coup and major human rights violation in Latin America in the past fifty years. What follows is a representative sampling of notable linkages:

- The three top Peruvian officers convicted in February 1994 of murdering nine university students and a professor are all SOA graduates.
- At least nineteen of the ranking Honduran officers linked to death squad Battalion 316 are SOA graduates, including battalion founder General Luis Alonso Discua.
- Former dictator General Manuel Noriega, a long-time Central Intelligence Agency operative currently serving forty years in a U.S. prison for drug trafficking, is an SOA graduate.
- SOA graduate Leopoldo Galtieri headed Argentina's military junta during a period in which thirty thousand people were killed or disappeared.
- More than 100 of the 246 Colombian officers cited for war crimes by an international human rights tribunal in 1993 are SOA graduates. Colombia's Lt. Col. Victor Bernal Castano was allowed to attend the School of the Americas in 1992 in order to escape a criminal investigation of his role in the "Fusagasuga" massacre of a peasant family.

Even this partial listing of atrocities associated with the U.S. Army School of the Americas and its graduates bolsters the case for closing the school. The example of

Guatemala offers another example of SOA complicity with torture and terror.

The case of Guatemala is compelling both because of the longevity of our country's ties to repressive rulers and because of recent horrific examples of abuse. As a *Washington Post* editorial noted, two recent deaths in Guatemala "are two among 150,000 or more in a dirty war that the United States helped along mightily by conspiring to oust the elected leftist leader in 1954."[16]

U.S. business interests, in this case the United Fruit Company, once again prevailed over the needs of common people, resulting in a CIA-led overthrow of a democratically elected, reformist government. Death and repression have stalked Guatemala ever since. In Guatemala, SOA-trained General Romeo Lucas García carried out a reign of terror during his dictatorial rule, 1978–82. General García was assisted by another SOA graduate, General Manuel Antonio Callejas y Callejas who, as chief of Guatemalan intelligence, oversaw the disappearance and assassination of thousands of political opponents. Callejas was not only trained at the SOA but he was selected as an elite member of the SOA Hall of Fame which honors distinguished alumni. His portrait hangs on a wall at SOA headquarters, along with portraits of former Bolivian dictator Hugo Banzer Suárez and other graduates.

As it turns out, the stream of blood flowing at the hands of Guatemalan graduates of the SOA runs deep and wide. One graphic case is that of SOA graduate General Hector Gramajo, a former Guatemalan Defense Minister. In a 1991 civil suit, a U.S. court found Gramajo responsible for the rape and torture of Diana Ortiz, a U.S. Ursuline nun. Sister Ortiz was working in San Miguel Acatán, a poor, rural area of Guatemala, teaching Mayan children to read, write and reflect on the Bible in the context of their Mayan culture. She was kidnapped on November 2, 1989, after having received numerous death threats. She recounted part of her story on Palm Sunday, in March 1996, as she stood in Lafayette Park, across from the White House, embarking on what was to be a long, silent vigil:

> I was abducted from the back yard of the Posada de Belén retreat center in Antigua by members of the Guatemalan security forces. They took me to a clandestine prison where I was tortured and raped repeatedly. My back and chest were burned more than 111 times with cigarettes. I was lowered into an open pit packed with human bodies — bodies of children, women, and men, some decapitated, some lying face up and caked with blood, some dead, some alive — and all swarming with rats.
>
> After hours of torture, I was returned to the room where the interrogation initially occurred. In this room I met Alejandro, a tall man of light complexion. As my torturers began to rape me again, they said to him, "Hey Alejandro, come and have some fun." They referred to him as their "boss." Alejandro cursed in unmistakable American English and ordered them to stop, since I was a North American nun and my disappearance had become public....
>
> Alejandro professed that he was concerned about the people of Guatemala and consequently was working to liberate them from communism. He kept telling me in his broken Spanish that he was sorry about what happened to me....I asked him what would happen to the other people I saw tortured. At this point, he switched to distinct, American English. He told me not to concern myself with them....He made it clear that he had been given the videotape and

> photographs that would incriminate me of crimes I had been forced to participate in. This was an obvious threat.
>
> The memories of what I experienced that November day haunt me even now. I can smell the decomposing bodies, disposed of in an open pit. I can see the blood gushing out of the woman's body as I thrust a small machete into her. For you see, I was handed a machete. Thinking it would be used against me, and at that point in my torture wanting to die, I did not resist. But my torturers put their hands onto the handle, on top of mine. And I had no choice. I was forced to use it against another human being. What I remember is blood gushing — spurting like a water fountain — and my screams lost in the cries of the woman. In spite of the memories of humiliation, I stand with the people of Guatemala. I demand the right to heal and to know the truth. I demand the right to a resurrection.[17]

There are many disturbing aspects to this story about the abduction, rape, and torture of Diana Ortiz for the "crime" of teaching Mayan children to read. First and foremost, of course, is the personal pain of her brutal experience. Sadly, however, her ordeal is unusual only because she is a U.S. nun, not because such brutal treatment is rare in Guatemala. "I begin my silent vigil for truth in front of the White House," she said, " — not a silence of complicity, but a silence of commemoration for those who have been tortured, assassinated, or disappeared in Guatemala in the last 30 years." Five weeks later she made another public statement that bridged Guatemala's past and present. "Over five weeks ago, I stood in Lafayette Park," she said, "along with other survivors of torture in Guatemala. The tulips were only slips of leaves; patches like open hands," she continued. "During two of the past five weeks I have fasted, losing 25 pounds. In Guatemala, approximately 10 people have been tortured since the tulips budded and bloomed."[18]

Diana Ortiz's ordeal is also troubling because it is part of the U.S. government's long and steady history of involvement with Guatemala's practitioners of terror. I remember a dozen or so years ago asking a Guatemalan leader-in-exile how much aid his country received from the United States. His answer illustrates U.S. foreign policy's bottom line in Guatemala and throughout much of Latin America, and it helps explain why the School of the Americas is associated with and has trained torturers. He smiled slightly, aware that I expected a specific number — in the millions of dollars, to be sure. "Enough," he said, simply. "Enough."

In Guatemala and elsewhere throughout the Americas, U.S. aid in one form or another was usually enough — sufficient, that is, to block any fundamental changes that might "threaten vital U.S. interests." There were of course a few costly miscalculations, such as when the U.S.-supported family dictatorship in Nicaragua unraveled after forty years, or when a socialist, Salvador Allende, was elected the president of Chile. Correcting these errors cost plenty and served to reinforce the desire of U.S. leaders to provide "enough."

Enough is a flexible category. Sometimes it means more money for the generals, sometimes less. Sometimes it means more training; at other times, less. Despite occasional philosophical or tactical differences with the forces of repression, the U.S. generally offered enough. If the human rights violations of Guatemalan officials proved embarrassing to a U.S. administration, for example, direct U.S. aid might fall and Israeli aid might rise, or military aid might fall and CIA operations increase. In combination, military aid when coupled with contact and training at the SOA and, most importantly, when linked to the CIA's constant presence, involvement, and funding always seemed to be enough. Guatemala's henchmen, who discarded bodies in open pits filled with rats, were adequately supplied to maintain control. Enough aid. Enough influence. Enough leverage. Enough support. "Vital interests" protected.

Related to the flexibility and longevity of U.S. ties to Guatemala's killing machine is the matter of secrecy, which casts a dark shadow over U.S.-Guatemala relations throughout a generally sorry history, and which hangs more specifically over the entire Diana Ortiz affair. At the School of the Americas and at the desks and dungeons of the Central Intelligence Agency, the U.S. conducts its policies by stealth. In Diana Ortiz's case, like so many others, official silence protects murderers and torturers and conceals U.S. training and support for the executioners. "Our own United States government," Sister Ortiz said on the day she began her silent vigil, "has been closely linked to the Guatemalan death squads, and has a great amount of detailed information about those of us who have survived as well as those who have perished. We need and demand this information," she continued, "so that we can heal our wounds, bury our dead, and carry on with our lives."[19]

The problem of official silence raises a particularly difficult issue concerning the case of the abduction and torture of Sister Ortiz, and the thousands of cases which it represents. The fact that we do not know the identity of Alejandro or his exact role within the Central Intelligence Agency (which in all likelihood is his employer) illustrates that secrecy and trickery are instruments of cover-up, pieces of intentional deception in service to murder.

State Department documents released in response to Diana Ortiz's fast and faithful protest reveal that the U.S.

Embassy in Guatemala "initiated a smear campaign against Ortiz immediately upon report of her abduction."[20] One document, dated March 19, 1990, urgently expresses the need to "close the loop on the issue of the North American named by Ortiz....The EMBASSY IS VERY SENSITIVE ON THIS ISSUE." Two completely blacked-out pages follow.[21]

Finally, Diana Ortiz's abduction, rape, and torture is another grisly example in which an SOA graduate is intimately involved in hideous deeds. General Hector Gramajo was ordered by a U.S. court to pay $47.5 million in damages based on the civil suit which held him responsible for Sister Ortiz's rape and torture. He ignored the order and blamed Ortiz's hundred-plus burn marks on a failed lesbian love affair.[22] Two years after his involvement in Ortiz's kidnapping and torture, General Gramajo delivered the commencement address to an audience of graduate officers for the Command and General Staff College of the School of the Americas. *The Bayonet,* Fort Benning's authorized newspaper, reported:

> Following the invocation, the guest speaker, retired Gen. Hector Gramajo from Guatemala, addressed the audience of graduate officers. Gramajo voiced concern for the continued vigilance in the Americas against communism and drug trafficking. Comparing the current state of communism to a dragon, Gramajo said the crumbling of the Berlin Wall signaled the beheading of the dragon; however, its tail is still poised to deliver a devastating blow to the countries of Latin America.[23]

Fittingly, it was General Gramajo who befriended another notorious Guatemalan graduate of the SOA, Col. Julio Roberto Alpirez. Alpirez was linked to the killing of Michael DeVine, an American innkeeper in Guatemala, and the torture and death of Efrain Bamaca Velasquez, a leader of the armed opposition, who was married to an American lawyer. Col. Alpirez, while on the CIA payroll, spent 1989 at the School of the Americas and then returned to Guatemala in 1990 where he continued working for the agency. A *Washington Post* editorial, "Our Man in Guatemala," offered this summary of Alpirez's ties to the CIA:

> The CIA, learning of these atrocities [the murders of DeVine and Bamaca], contained and covered up the relevant information, ostensibly to protect "sources and methods." Officials at the State Department and National Security Council kept the full story from the American wife. Word finally got out only as a result of disclosures by Rep. Robert Torricelli.
>
> It defies credulity that, at this late date in the United States Central American involvement, the CIA could still be recruiting killers of the sort that have made Guatemala's the region's bloodiest army. To hire an informant is one thing. To condone his criminality, by doing nothing to bring him to justice after two murders, is to lend official American approval — on the level where it counts most — to the Guatemalan military's criminal habit.[24]

General Gramajo's arguments in defense of Col. Alpirez are, as we shall see, about as weak as those used by defenders of the School of the Americas. He referred to Col. Alpirez who, according to U.S. Representative Robert G. Torricelli, was an SOA graduate, a well-paid secret operative for the Central Intelligence Agency, and a murdering spy, as "a soldier above all." Alpirez, according to General Gramajo, had so distinguished himself as a young cadet as to be rewarded with a trip to the School of the Americas for further training. "I am sure he is a good officer," General Gramajo said of Alpirez. "He does not drink. He does not argue. He is a good father. He is the kind of officer who you would want under your command."[25] U.S. foreign policy planners and School of the Americas trainers apparently agree.

— ❖ —

Despite overwhelming evidence defenders of the SOA insist that the School is doing good work and should not be judged by the behavior of a few problem students. This "few bad apples among many good apples" argument obviously spoils under the weight of the evidence provided above. It will be discussed more fully, however, in the following chapter. Here I have offered numerous and compelling examples linking the U.S. Army School of the Americas and its graduates with dictatorship, human rights abuses, torture, and terror. It should be remembered that this is by no means *all* the evidence that could be cited. It is a *sampling* of evidence which I hope makes a compelling case for closing the school.

The evidence presented above is disturbing. It makes many people sad or angry. It prompts some to action. Almost universally, however, it leaves people wondering. Why, they ask, has the School of the Americas trained so many graduates who end up being assassins, torturers, and human rights violators? And why, despite this record of abuse, does the school have many defenders and remain open? An honest probe of these questions, as we shall see, takes us far beyond the theory of a "few bad apples." It strikes at the heart of the ideology which drives U.S. foreign policy. It is to these questions that we now turn.

Chapter 3

Lies and Consequences

SOA Defenders

> They boast about the fact that 10 separate heads of state throughout Latin America were graduates of the School of the Americas. Not one of them was elected through a democratic election, and in many cases they actually overthrew the civilian governments that brought them into power. They tell us now that the school [is] changing, but we know and understand... that the school is continuing the kind of *modus operandi* that left us with the legacy of being associated with some of the worst human rights abusers on the face of the planet.
>
> — REPRESENTATIVE JOSEPH KENNEDY[26]

Despite the evidence presented in the previous chapter, the SOA has its supporters. Their arguments on behalf of the SOA fall into four main areas.

- First, as was previously mentioned, supporters say it is unfair to blame the School of the Americas for the conduct of a few "bad apples."
- Second, SOA supporters point to many benefits which flow from exposing Latin American soldiers to our culture, values, and democratic way of life.

- Third, SOA supporters highlight positive curricula offered at the School of the Americas, including courses which promote human rights, nation building, respect for civilian authority, professionalization of the armed forces, and democracy.
- Finally, SOA supporters intimate that vital U.S. national security needs have been and continue to be met through the activities of the School of the Americas.

The previous chapter ended with several troubling questions. Why has the School of the Americas trained so many graduates who end up being assassins, torturers, and human rights violators? And why, despite this record of abuse, does the school have prominent defenders and remain open? In order to begin to probe answers to these and other disturbing questions, the arguments used by SOA supporters need to be scrutinized.

The most common defense of the School of the Americas is the "occasional-bad-apple-among-many-good-apples" argument. When you purchase a barrel of apples you don't throw out the good ones because you find a couple of bad ones with worms, SOA supporters say. You most certainly don't cast blame on the barrel in which mostly good apples are found. The *Columbus Ledger-Enquirer* argues on its editorial page that the SOA should not be held responsible for the "behavior of all of its graduates, including those who commit human rights abuses." "You can find," the editorial continues, "criminals of every ilk who graduated from Harvard, Yale, Princeton," but "no one advocates closing those institutions because of the crimes of some of their graduates." In a similar vein, Col. José Alvarez, SOA's commandant, asks, "Does the Wharton School take the blame for Michael Milken?"[27] Douglas Waller, who includes Col. Alvarez's question in a *News-*

week article depicting the School of the Americas' long list of unsavory graduates, notes: "If the Wharton School had this many Milkens, maybe it would."[28]

Given the weight of the evidence presented in the previous chapter, the "bad apple" argument isn't very convincing: two of three officers cited in the assassination of Archbishop Romero; three of five officers cited in the rape and murder of the four U.S. churchwomen; ten of twelve officers cited as responsible for the massacre of nine hundred civilians at El Mozote; nineteen of twenty-six officers cited in the murder of the Jesuit priests; over two-thirds of the more than sixty officers cited for the worst atrocities in El Salvador's brutal war — all SOA graduates, all exercising leadership, all abusing power.

"SOA defenders say that the school's reputation, any more than that of other institutions, should not be tainted by the wrongdoings of a relatively small number of graduates," notes an editorial in the *Cleveland Plain Dealer,* Ohio's largest paper. "But the SOA's best known products have shared a distressing tendency to show up as dictators or as leaders or members of death squads."[29] Despite its appeal, the "bad apple" argument spoils under the weight of the evidence suggesting that the school's disproportionate number of bad, if not notorious, apples reveals it to be rotten to the core.

The second line of defense used by SOA supporters is that the exposure to American values which Latin American students receive while attending the School of the Americas makes them better officers and soldiers. According to its defenders, a central goal of the School of the Americas is to immerse students in American values and culture. Putting aside the thorny issue and possibility that SOA graduates may have learned how to terrorize civilians and to torture while studying at the School of the

Americas, an issue to which we will return, it is at best a questionable assumption that exposing Latin American soldiers to American values and our way of life will encourage behaviors that are conducive to democracy.

Saturating Latin American officers with the perks associated with U.S. affluence may itself be a problematic goal, no matter how well-intentioned. As one critic within Fort Benning notes, taking officers to Atlanta Braves baseball games and Disney World "shows them the good life, not democracy. They go home," he continues, "thinking that if their army stays in power they can continue the perks they experience here."[30] We might also ask, what lessons Latin American soldiers learn when they are invited to attend the School of the Americas and receive the perks listed above even *after they have committed horrendous abuses?*

A third argument used by SOA supporters is that the courses offered at the School of the Americas promote democracy and human rights. Far from being responsible for abuses, supporters say, the School of the Americas is a defender of the very values and concerns that its critics hold dear. If these critics of the school really cared about human rights and democracy, this line of reasoning suggests, then they should be ardent defenders of the school. Joe Reeder, Undersecretary of the Army, notes:

> SOA instruction focuses on the role of the military professional in a democratic society. It is a requirement of the School that every course, regardless of subject or length, include formal instruction emphasizing the sanctity of human rights and the proper role of the military in a democratic society.[31]

Such courses and the values and priorities they reflect have yielded dramatic results, according to Democrat Sam Nunn, who at the time was Georgia's senior senator. Speaking at a press conference at Fort Benning called by SOA officials to defend the School of the Americas against critics, Nunn said:

> Democracy after democracy has emerged, because of American values and human rights exported to Latin America. Our relations with our neighbors in Latin America are stronger because of the School of the Americas.[32]

In Nunn's estimation the fact that only two-thirds of the officers named for the worst atrocities in El Salvador were SOA graduates is apparently something of a success story. He is not alone. "The conflict waged in El Salvador," writes Joseph C. Leuer in a January 1996 paper produced by the School of the Americas to counter the growing tide of criticism, "highlights the positive outcomes of engaging the armed forces and effecting real change within the institution. SOA played a critical role in the development and the professionalization of the Salvadoran Armed Forces."[33] Leuer cites a letter from the former U.S. Ambassador to El Salvador, Edwin Corr, to bolster his position that SOA critics have it backwards when they base their reproaches of the School of the Americas on the conduct of SOA graduates in El Salvador:

> El Salvador had a long tradition of repression and human rights violations. Training at SOA did not cause nor intensify human rights violations. . . . In the area of human rights, the influence of SOA on the El Salvadoran military students was . . . to curb, reduce and change ingrained patterns of behavior abusive to the citizenry. SOA training did not intentionally nor inadvertently encourage violations of citizens' rights.

> ...The Salvadoran military underwent an amazing metamorphosis, in the midst of a bloody civil war....[The change] was made possible by the access and relationships that U.S. civilian and military officials had with Salvadoran military officers....This access and these relationships were greatly enhanced by the training and democratic experiences Salvadoran soldiers had gained at SOA.[34]

Father Roy Bourgeois, who as a priest worked in the slums on the outskirts of La Paz, Bolivia, during the brutal dictatorship of SOA-graduate Hugo Banzer Suárez, sees the school and the conduct of its graduates differently. He witnessed human rights atrocities against tin miners, factory workers, university students, professors, and religious workers during Banzer's reign of terror. After visiting the prisons and seeing widespread evidence of torture he spoke out against the dictatorship during a visit to the United States. As a result, Bolivian authorities prevented his return.

It would be difficult to find a more compelling and revealing portrait of contrasting worlds than this. Picture Father Bourgeois, who experienced Banzer's repression, doing jail time on several occasions after protests calling attention to the School of the Americas and the unsavory conduct of its graduates. Now picture Senator Sam Nunn standing within a stone's throw of the School of the Americas' Hall of Fame, where a portrait of General Banzer hangs on the stairwell leading to the SOA Commandant's office. Standing in the shadow of numerous dictators, Senator Nunn lauds the School of the Americas for its export of democracy, American values, and human rights.

Father Bourgeois points out what is missing from Joe Reeder's glowing statements about SOA curricula, from

Senator Nunn's and former Ambassador Corr's songs of praise, and from Joseph C. Leuer's whitewash of SOA abuses:

> The SOA is a combat school. Courses include commando operations, sniper training, how to lay land mines, psychological operations, and interrogation techniques. Soldiers are trained in "Low Intensity Conflict," a cynical and relatively new strategy for maintaining U.S. military influence south of our borders, without using (or losing) large numbers of U.S. troops. Instead, soldiers from Latin America and the Caribbean are trained in "dirty little war" techniques by U.S. personnel.[35]

The central curriculum of the School of the Americas since its founding in 1946 has been counterinsurgency, with the predictable results detailed in the previous chapter. Exactly how the School of the Americas' emphasis on counterinsurgency helps explain the association of its graduates with terror and torture will be detailed in later chapters. I mention it here because supporters of the school state or imply that the goals of the school are to promote democracy, to enhance human rights, and to professionalize military forces. This is simply not true. Supporters of the School of the Americas often base their defense on recent changes in SOA curricula. This is questionable for four reasons. First, supporters minimize SOA's deadly track record of abuse. Georgia Congressman Sanford Bishop fought efforts to close the School of the Americas with the following argument:

> The school was developed to train and teach Latin American militaries how to defend against subversion techniques from the Soviet Union and Cuba. How-

> ever, as the cold war began to end, the School of the Americas began to adopt a new curriculum. The new emphasis began to focus on the role of military professionalism in a democratic government. The guiding principle of the school now is to provide professional service subordinate to civilian control by democratically elected governments. Training at the school is focused on effective response to drug trafficking, natural disasters, and respect for human rights.[36]

Congressman Bishop has a remarkable capacity to minimize the past while painting a rosy picture of the present.

Second, as mentioned previously, present classes on human rights are often read back into history as if human rights training has always been an important component in U.S. training at the School of the Americas or as if it can somehow erase the past. As Congressman Bishop acknowledges above, it was only as "the cold war began to end" that "the School of the Americas began to adopt a new curriculum." Third, references to SOA human rights courses presently taught are used to suggest that human rights training today occupies center stage at the School of the Americas. Finally, the human rights courses offered at the School are portrayed as dynamic and effective means by which the United States spreads American values and human rights throughout the Americas. None of these positions has merit.

Charles T. Call, an associate for U.S. hemispheric security policy at the Washington Office on Latin America, was the first human rights advocate to be invited to give a lecture at the School of the Americas. Call reflected on his experience at the SOA in an article in the *Miami Herald:*

> Conscious of its reputation [of links to human rights abusers], the school has recently given more attention

> to human rights in its curriculum. In the last three years, it has formally incorporated human rights into its classes and practice exercises....
>
> Unfortunately, I found that these changes are not much more than a facelift.... Several instructors, I found, are from countries with appalling human rights records.... Indeed, much of the training at the school is done by officers from Latin American militaries, which have strongly resisted increased civilian control and accountability. Yet the Defense Department invites officers from these militaries to serve as teachers and role models.
>
> And in spite of the new language about human rights and democracy, U.S. trainers appear to pay only lip service to these goals. Col. José Feliciano, then-commandant of the school,... had displayed in his office a 1991 letter and gift sword from Gen. Augusto Pinochet, the former Chilean dictator who became a model of harsh repression.... Despite the end of the Decade of the Dictators, our military seems reluctant to break the symbols and attitudes of that era. Even more distressing, I found that the United States continues to invite soldiers accused of gross human rights violations to the school.... Without [substantial changes], the Clinton administration will continue the United States' Cold War habit of bolstering non-democratic and abusive militaries abroad.[37]

Call told a *Newsweek* reporter on another occasion that Latin American instructors teaching at the School were particularly hostile: "All they wanted to do was bash human-rights groups."[38] This might help to explain the disappointing results of one human rights training exercise carried out at the school in which Latin American soldiers are taught to use restraint when dealing with priests and catechists. Soldiers are to retake control of a mock town that has fallen into the hands of rebels. During the exercise, however, the priest and catechists were either killed or roughed up seventy-five percent of the time. The frequent death of the priest was treated as a joke which circulated around the School.[39]

Call's assessment that changes in curriculum at the School of the Americas are more cosmetic than real is shared by Retired U.S. Army Major Joseph Blair who taught logistics at the School of the Americas from 1986 to 1989. Tim McCarthy interviewed Blair as part of a story on the School of the Americas for *NCR*. Blair indicated that four-hour blocks of instruction on human rights awareness were offered at the School, which he described as "a bunch of bullshit." Blair attended one class taught by a Chilean army officer whom he called "a Pinochet thug." According to Blair, most Latin American soldiers regard the human rights training as a joke. Human rights are associated with subversives, primarily the Catholic church and other groups such as Amnesty International.[40]

In an article published on the opinion page of the *Columbus Ledger-Enquirer,* Major Blair criticized the School of the Americas and called for its closure:

> In three years at the school, I never heard of such lofty goals as promoting freedom, democracy, or human rights. Latin American military personnel came to Columbus for economic gains, opportunities to purchase quality consumer goods with import tariff exemptions in their countries, and for free return transportation at U.S. taxpayers' expense. American faculty members readily accepted all forms of military dictatorships in Latin America and frequently

> conversed about future personal opportunities to visit their new "friends" when they ascended to positions of military or dictatorial power some day. U.S. officers routinely used their positions for junket-style visits to Latin America and to seek personal gains for their forthcoming retirement years. . . . With the end of the Cold War and the reduction in our Department of Defense, I find it hard to believe that anyone can justify the continuance of the School of the Americas on any political or military grounds.[41]

If the School of the Americas has been attempting to teach values other than those which spawn torture and abuse, then the record would indicate that the school has failed miserably, Nunn's optimistic assessment notwithstanding. According to Representative Joseph Kennedy, who has introduced legislation to close the SOA, by "any reasonable standard, the extensive record of abuse by the school's graduates demonstrates that it has failed in one of its central missions — teaching respect for human rights and civilian authority. Through the school, the United States continues to be associated with those abuses."[42]

The fourth line of defense used by supporters of the School of the Americas is that the SOA plays a critical role in serving vital U.S. national security interests. This justification for the SOA might appear at first glance to be as weak as the first three. Our first impulse may be to dismiss it as the propaganda of those with a vested interest in keeping the School of the Americas open. But what if "vital national security interests" were/are at stake, at least in the minds of U.S. foreign policy planners? What if the training offered at the School of the Americas is designed to defend these perceived interests? What if Congressman Bishop, cited above, is correct when he says that the "school was developed to train and teach Latin American militaries how to defend against subversion techniques from the Soviet Union and Cuba"? If, in the minds of U.S. policy makers, the interests being defended are important enough and the enemies subversive enough, then we may have uncovered compelling reasons why torture and terror could at times be taught or encouraged at the School of the Americas.

Beyond the officially stated goals of promoting democracy and professionalizing armies, SOA defenders note that protecting vital U.S. interests depends on cultivating ties with military leaders throughout Latin America. "If Americans want a say in how nations conduct themselves, they have to have a seat at the table," SOA commandant Col. José Alvarez says. "What this school does is give you a seat at the table with the armies of Latin America."[43]

According to Alvarez's logic, the School of the Americas serves as a vital link to military groups throughout Latin America who, in addition to the power of their weapons, are often major political and economic players. "I strongly believe," writes former Ambassador Corr, "that SOA is a valuable tool for assuring access to the military in Latin America, which remains an important political force in the post-cold war era."[44] SOA Chief of Staff Lt. Col. Kohn Bastone defends the School of the Americas precisely on these grounds. "SOA alumni have attained positions of prominence," he boasts, "which include 10 presidents of the republic, 15 ministers of national departments, 23 ministers of defense" and many others.[45]

Bastone's assessment is historically misleading: the ten "presidents" he mentions were dictators who trained at the School of the Americas and then illegally assumed power. Bastone, however, makes the important point that the School of the Americas has helped the United States train, gain access to, or place people in positions of power and influence throughout the Americas. His defense of the school, therefore, goes beyond those discussed previously, including the "bad apple" argument or justifications based on either the export of American values or the enhancement of human rights and democracy based on SOA's enlightened curricula. Bastone speaks instead of power and influence. What his assessment ignores, however, is the overwhelming body of evidence that SOA graduates generally, and perhaps more specifically those who Bastone says "have attained positions of prominence," have trained at the School of the Americas, returned to their countries, and carried out human rights atrocities. Lt. Col. Bastone, Edwin Corr, and Col. Alvarez seem to share the same rose-colored glasses and wear the same blinders.

Alvarez, quoted above, speaks of the leverage which comes from sitting at a common table with Latin American military leaders. Retired Vice Admiral T. J. Kilcline, president of the Retired Officers Association, concurs:

> The impact of the school has truly been significant. Not only has the education been most helpful for our Latin American neighbors, but the contact with Americans and the positive attitudes of the American military personnel they met and got to know while at Fort Benning was the basis for friendship and understanding between individuals which translates to better relationships among our countries.[46]

This image of the SOA as a table where Latin American officers chat amiably with their U.S. counterparts, soaking up our values and culture in the process, is highly misleading. This idyllic portrayal is little more than fantasy.

For decades trainers at the School of the Americas have been serving up vast banquets, with most of the food for thought having to do with combating internal enemies, fighting communism, and undermining subversive movements. No wonder many such meals end with bitter fruits, including massacres, torture, and dramatic human rights abuses.

Father Roy Bourgeois of SOA Watch interviewed several former SOA graduates for an hour-long documentary titled *Inside the School of Assassins.* One of the graduates, who spoke on film on condition that his face and name not be used, said:

> The school was always a front for other special operations, covert operations. They would bring people from the streets [of Panama City] into the base and the experts would train us on how to obtain information through torture. We were trained to torture human beings. They had a medical physician, a U.S. medical physician which I remember very well, who was dressed in green fatigues, who would teach the students... [about] the nerve endings of the body. He would show them where to torture, where and where not, where you wouldn't kill the individual.[47]

A detailed investigation by Gary Cohn and Ginger Thompson of the *Baltimore Sun* confirms allegations that the United States was intimately connected to death squads and torturers. According to their report the "CIA was instrumental in training and equipping Battalion 316," a secret army unit that was home to Honduran death squads:

> The intelligence unit, known as Battalion 316, used shock and suffocation devices in interrogations. Prisoners often were kept naked and, when no longer useful, killed and buried in unmarked graves. Newly declassified documents and other sources show that

> the CIA and the U.S. Embassy knew of numerous crimes, including murder and torture, yet continued to support Battalion 316 and collaborate with its leaders.[48]

José Valle, a former School of the Americas graduate, a member of Battalion 316, and an admitted torturer, said torturing was "a job, something I did to give food to my kids."[49] Valle told Father Bourgeois that he took "a course in intelligence at the School of the Americas" in which he saw "a lot of videos which showed the type of interrogation and torture they used in Vietnam. . . . Although many people refuse to accept it," Valle said, "all this is organized by the U.S. government."[50]

In 1996 the Pentagon was finally forced to admit that training manuals used at the School of the Americas instructed Latin American officers in the art of execution and torture. We will look at the content of these manuals in more detail later. An editorial in the *Boston Globe*, titled "Lessons in Terror," gives a flavor of their content:

> Murder, extortion, torture — those are some of the lessons the US Army taught Latin American officers at the notorious School of the Americas in Columbus, GA. Recent revelations that the Pentagon trained police and military leaders in committing blatant atrocities describe a program that is beyond redemption.[51]

Lt. Col. Bastone and other supporters of the School of the Americas suffer from historical amnesia. The prominent positions held by SOA graduates, of which Bastone boasts, should be, according to many critics of the school, a reason for shame. Tim McCarthy, writing in the *Na-*

tional Catholic Reporter, quoted Bastone (above) and then countered his logic:

> Apart from Noriega, those esteemed graduates include Gen. Hugo Banzer Suárez, Bolivian dictator 1971–78, who brutally suppressed progressive church workers and striking tin miners; Gen. Romeo Lucas García, Guatemalan dictator 1978–82, whose bloody reign saw at least 5,000 political murders and up to 25,000 civilian deaths at the hands of the military; Gen. Policarpo Paz García, corrupt dictator of Honduras, 1980–82; and Gen. Juan Rafael Bustillo, former Salvadoran air force chief, who was cited in the 1993 U.N. Truth Commission report for helping to plan and cover up the Jesuit massacres.[52]

The General Bustillo-headed air force, it should be remembered, produced and distributed a leaflet shortly before the murder of the Jesuits. It read:

> Salvadoran Patriot! You have the . . . right to defend your life and property. If in order to do that you must kill FMLN terrorists as well as their 'internationalist' allies, do it. . . . Let's destroy them. Let's finish them off. With God, reason, and might, we shall conquer.

Not to be outdone, U.S.-trained soldiers from San Salvador's First Infantry Brigade circled the office of the Catholic archdiocese shortly after the Jesuit murders. From a military sound truck they shouted, "Ignacio Ellacuría and Ignacio Martín-Baró have already fallen and we will continue murdering communists."[53]

Let me summarize. In the previous chapter I presented a lengthy yet partial list of SOA graduates involved in human rights atrocities. It demonstrates, convincingly I think, that the "bad apple" argument used as the first line of defense by SOA supporters can't stand the weight of careful scrutiny. The second and third lines of defense imply that beyond a few "bad apples" the school can celebrate many notable achievements due to successful cultural immersion experiences and the strength of the SOA's enlightened curricula. These defenses also fail to hold up under careful scrutiny. Experiencing the good life may encourage military leaders to hold onto power because of the perks they enjoy. And recent changes in SOA curricula cannot cover up past abuses or the fact that such changes are more cosmetic than real. In other words, these defenses employed by SOA supporters come crashing down under the weight and volume of atrocities associated with SOA graduates.

An editorial in the *Syracuse Post-Standard* states forcefully: "Supporters claim the SOA played a role in bringing democracy to Latin America over the past decade. Hogwash. The evidence," the editorial continues, "suggests a much greater role for the SOA in the decades of military terror that preceded democratization."[54]

Exposing the weaknesses of the first three defenses of the School of the Americas forces us seriously to assess the fourth, namely, that the school helps defend vital national security interests. As much as we might want to dismiss this point of view, we would be wrong to do so. If the "bad apple" defense doesn't hold, then we have to look elsewhere if we are to understand the undeniable and extensive link between the School of the Americas, SOA graduates, and torture.

Just how and why U.S. foreign policy and the School of the Americas became so closely aligned with dictators and torturers is the subject of the next two chapters. We will see that the United States opened the school following World War II in order to shore up ties to the militaries throughout Latin America. The school's primary objective was to assist the military and police forces of a given country to maintain or reestablish control in any given environment. Over time the main focus of training became counterinsurgency with a focus on internal enemies. The school did not set out to feed sadistic impulses, but in a twisted yet logical way managing torture and repression became the focus of human rights training. The idea was to maintain stability while employing the *least* amount of repressive force or violence. Torturing political opponents was considered necessary in one context and counterproductive in another. Human rights classes tried to help officers and soldiers learn the difference. In actual practice, however, support for torture, terror, and dictatorial rule became commonplace because enemies were vilified, stability was desired, and justice was a non-existent goal.

From the vantage point of SOA trainers, torture is neither good nor bad in itself. It is, however, sometimes necessary. Like any other policy option, torture is employed when circumstances require it. This view of torture as one of many tactical options is, unfortunately, part of the tragic history of the School of the Americas. I find it particularly troubling that Salvadoran officers cited by the Truth Commission have attended the SOA *after* their involvement in deadly atrocities became known. It hardly seems a judgment befitting their crimes when known human rights abusers receive weekend trips to Disney World, attend Atlanta Braves baseball games, and enjoy the benefits of the recent $30 million renovation of school headquarters and Latin American bachelor officer quarters all at U.S. taxpayer expense. According to the Pentagon, SOA operating expenses are $18.4 million a year.

More troubling still, and lending credence to the view that human rights and other abuses linked with SOA graduates aren't coincidental but are instead linked to the goals, training, and ideological backdrop of the school itself, is the fact that officers such as Retired General Hector Gramajo and others cited by the Truth Commission have returned to the SOA as guest speakers or as guest instructors even *after* their involvement in massacres and human rights violations became public knowledge.

The "bad apple" defense can't hold, nor can the fantasy of success that flies in the face of so many atrocities associated with the School of the Americas and its graduates. We are left with a disturbing conclusion: the conduct of the School of the Americas and SOA graduates, including numerous linkages to human rights abuses, is a result of policy. As an SOA brochure states succinctly, the SOA is "an implement of foreign policy."[55] The foreign policy behind the abhorrent practices of the School of the Americas and its graduates is the subject of the next two chapters.

Chapter 4

History 101 Revisited

> It is now clear that we are facing an implacable enemy whose avowed objective is world domination. ...There are no rules in such a game. Hitherto accepted norms of human conduct do not apply....If the United States is to survive, long-standing American concepts of fair play must be reconsidered....We must learn to subvert, sabotage and destroy our enemies by more clever, sophisticated, more effective methods than those used against us.
>
> —The Hoover Commission[56]

In order to discover how the School of the Americas and its graduates fit into U.S. foreign policy, and why both the school and its alumni are implicated in human rights violations, it is necessary to take a detour into history at least as far back as the end of World War II. The number one priority of the United States following the war was to rebuild a devastated Europe. The collective weight of enormous humanitarian needs, profound concerns about the possible direction of progressive European political movements, and the need for a rebuilt Europe to serve as a market for otherwise surplus U.S. production spawned the Marshall Plan. Stability in Europe and constructing a post-war international economy under U.S. control, according to U.S. policy makers, depended on massive economic aid.

As strange as it might seem, many of the political alliances which the United States built throughout Europe following World War II were not with those who had led the resistance to Nazi occupation, on whose behalf we were fighting, but with more conservative forces, including in many cases those who had collaborated with the Germans. It is important that we acknowledge this because it helps ground our understanding of why and how, outside Europe, the U.S. relied upon repressive forces to protect its perceived interests. This in turn helps explain the deadly legacy of the School of the Americas.

Preoccupation with Europe had profound implications for Central and Latin American countries and other nations outside Europe. Most notably, a European-centered policy enhanced the power and position of the military sector within the so-called Third World. Whereas U.S. policy makers set out to stabilize Europe through an influx of massive economic aid, they turned throughout the Third World to military leaders to establish and maintain stability and to protect vital U.S. interests (a euphemism for U.S. corporate investments).

The treasurer of Standard Oil of New Jersey stated the logic behind U.S. policy in a speech before the National Foreign Trade Convention in 1946, the same year the U.S. Army School of the Americas opened in Panama:

> American private enterprise is confronted with this choice; it may strike out and save its position all over the world, or sit by and witness its own funeral.... We must set the pace and assume the responsibility of the majority stockholder in this corporation known as the world.... This is a permanent obligation.... Our foreign policy will be more concerned with the safety and stability of our foreign investments than ever before.[57]

The mythology of U.S. support for democracy is so strong that it is often difficult to challenge, even when the historical record speaks otherwise. In 1948, George Kennan, who at the time headed the State Department's planning staff and who following World War II was arguably the most important architect of U.S. foreign policy, stated the underlying assumptions guiding U.S. policy:

> We have about 50 percent of the world's wealth, but only 6.3 percent of its population.... In this situation, we cannot fail to be the object of envy and resentment. Our real task in the coming period is to devise a pattern of relationships which will permit us to maintain this position of disparity without positive detriment to our national security. To do so we have to dispense with all sentimentality and day-dreaming; and our attention will have to be concentrated everywhere on our immediate national objectives. We need not deceive ourselves that we can afford today the luxury of altruism and world-benefaction.

Speaking specifically to Asia, but with relevance for Central and Latin America, Kennan added:

> We should cease to talk about vague and... unreal objectives such as human rights, the raising of living standards and democratization. The day is not far off when we are going to have to deal in straight power concepts. The less we are hampered by idealistic slogans, the better.[58]

Although preoccupied with war-ravaged Europe, the United States had "vital interests" throughout the Third World where social conditions such as hunger, poverty, and inequality were feeding deep impulses for social change. In this context, the U.S. put its considerable ideological and financial weight and influence behind often repressive militaries. The goal of U.S. foreign policy, as Kennan stated clearly, wasn't democracy, freedom, development, or human rights. It was stability.

Joseph C. Leuer, in his contradiction-laden defense of the School of the Americas, describes "the School's role as a forum to establish stability through military-to-military engagement."[59] Latin American officers trained at the School of the Americas were instruments of U.S. foreign policy. They were equipped to establish and maintain the stable investment climate favored by U.S. political and business leaders.

Quite predictably, U.S. foreign policy, which sought stability in the absence of democracy, freedom, development and human rights relied increasingly upon force and intimidation. Already predisposed to rely upon the generals to defend their interests, U.S. foreign policy planners found the School of the Americas to be a valuable asset as problems of hunger, poverty, and social inequality deepened during the rule of the generals. "The U.S. public," Leuer writes, "has difficulty understanding past U.S. support of sometimes authoritarian regimes in view of reported human rights abuses that have been attributed to them by international human rights organizations. However, past governmental support for many oligarchic Latin American regimes was accepted as necessary...."[60] At one point Leuer claims such support was necessary "to stimulate the transition to participatory democratic governmental structures in a competitive bipolar world."[61] Elsewhere he acknowledges more honestly that, following World War II, "foreign aid programs, specifically the Mutual Securities Act of 1951, began to tie foreign development aid directly to military aid and anti-Marxist allegiances. At the time *the military was perceived as the only stable force which could achieve the U.S. goal* of denying access to government by the revolutionary thinkers" (emphasis added).[62] Ironically, Leuer names and then passes over the brutal contradictions at the heart of U.S. policy.

> U.S. analysts believed there were three major factors which would allow for communist expansion in Latin America: (1) Latin American resentment of U.S. intervention in the Americas, (2) Neoclassical development schemes imposed on Latin American governments by large U.S.-controlled multinational companies, and (3) U.S. support of Latin American elites tied directly to the repressive military structure and the United States.[63]

No wonder the School of the Americas, established in 1946 to promote stability throughout the region, became known by other names: "School of Coups," "School of Assassins," and "School of Dictators." As the *Atlanta Constitution* notes, in an editorial urging that the School of the Americas be shut down:

> Over the years, the school developed a reputation that was the opposite of stability. So many of its attendees played leading roles in overthrowing governments — Panama, Ecuador, Bolivia, Peru and Argentina — that their alma mater came to be known as "la escuela de golpas" (the school for coups).
>
> A decade ago it acquired a more ominous nickname, the School for Assassins, from a Panamanian newspaper at the time Panama severed its links with the school. The label was to call attention to the unhappy penchant of the school's alumni to turn up as suspected death-squad officers throughout Central America.[64]

A year after the opening of the U.S. Army School of the Americas in Panama, the U.S. Congress passed the National Security Act of 1947. This act created the Central Intelligence Agency and the National Security Council. The ethical grounding for these agencies, which have since been linked to numerous acts of deception, destabilization, and terror, was the belief that the United States must use *any means* necessary to defend its vital interests.

Within the parameters of this logic, the CIA and its extensive network of assets, contacts, and agents became a sort of presidential hit squad sent out in the name of national security. The mission, in the words of the Hoover Commission cited at the beginning of this chapter, was "to subvert, sabotage and destroy our enemies." The means used to achieve these objectives not only violated "hitherto accepted norms of human conduct," as the evidence against the School of the Americas makes clear, but they often circumvented the law, the will of Congress, and the conscience and moral sensibility of the people. Despicable means and goals were shrouded in secrecy.

In *The Iran Contra Connection,* Peter Dale Scott, Jonathan Marshall, and Jane Hunter write of the origins of U.S. covert capabilities:

> From their inception to the present, many CIA operations have been covert, not just to deceive foreign populations, but at least partly because they were *designed* to violate U.S. statutes and Congressional will. A relevant example is the so-called "Defection Program" authorized in 1947 (by National Security Council Intelligence Directive 4, a document still withheld in full). Despite explicit Congressional prohibitions, this program was designed to bring Nazi agents, some of them wanted war criminals, to this

country, to develop the covert capability of the United States.[65]

Once the United States successfully established its own intelligence and covert operations capability, it set out to assist friendly military regimes throughout the Americas in setting up similar structures. The result, as José Comblin documents in his book *The Church and the National Security State,* was that throughout Latin America state power was exercised through military leaders and institutions.

Comblin called such military-dominated nations national security states. Such states, he said, set legal or functional limits on constitutional authority; justified human rights and other abuses committed by agents of the state by appealing to higher values or defense of the state itself; sought national unity based on attacks against external or internal, usually communist, enemies; and positioned ultimate state power in military hands.[66]

In my own work *Brave New World Order,* I adapted Comblin's original categories in light of my own experience in Central America in the mid-1980s. Readers will find more details there. Here I want only to list the seven key features of national security state ideology which shed light on the role of the School of the Americas and its graduates in implementing policies of terror and torture:

1. In a national security state the military is the highest authority.

2. In a national security state political democracy and democratic elections are viewed with suspicion, contempt, or in terms of political expediency. National security states may maintain an appearance of democracy. Ultimate power, however, rests within the military or a broader national security establishment. The generals sometimes utilized elections as a cover for their own de facto rule. U.S. and

Latin American military leaders concluded a secret defense plan at a series of meetings in Argentina in 1987 by stating their opposition to a new wave of military coups. Their preferred choice was "a permanent state of military control over civilian government, while still preserving formal democracy."[67]

3. In a national security state the military not only rules with the persuasion of guns, it wields substantial political and economic clout. An ideological foundation of many national security states is the conviction that "freedom" and "development" depend on the concentration of capital, either within the private sector or within institutions of the state itself.

4. A national security state is obsessed with enemies and it is the fight against enemies which gives legitimacy to the state's military defenders.

5. National security states define their enemies as ruthless and cunning. Dehumanization of enemies legitimizes a national security state's strategy for warfare employing "any means necessary." Terror, torture, and other human rights violations are viewed as legitimate tactics in a war against "implacable" enemies.

6. National security states often use intimidation and secrecy to restrict public debate, limit popular participation, and cover up abuses.

7. Finally, national security states expect the church to mobilize financial, ideological, and theological resources in service to the state. Church workers who violate this informal agreement by organizing with the poor, for example, are often defined as enemies and targeted for repression.[68]

Full-fledged national security states, and the ideology which guided their function and formation, did not emerge fully in Latin America until after the anti-communist hysteria which followed the Cuban revolution in 1959. But their seeds were planted and watered by U.S. policies much earlier.

The quote with which I began this chapter is taken from a top secret report prepared at the request of the White House in 1954. That was the same year the CIA carried out its notorious coup in Guatemala setting off more than forty years of almost uninterrupted terror. It was also the same year that President Eisenhower presented the Legion of Merit to two Latin American dictators—Pérez Jiménez of Venezuela (for his "spirit of friendship and cooperation" and his "sound foreign investment policies") and Manuel Odría of Peru.

The content and recommendations of the Hoover Commission; the bestowal of Legion of Merit awards to dictators; the CIA-led overthrow of a democratically elected government in Guatemala; Kennan's dictate that we could not afford the luxury of altruism and should therefore cease to talk about vague and unreal objectives such as human rights, the raising of living standards, and democracy; "Defection Directives" bringing Nazi war criminals to the United States to help establish intelligence operations; the export of national security state ideology; alliances with military despots; and training of torturers at the U.S. School of the Americas are all pieces of a hidden history unknown to most Americans deeply socialized in the comforting rhetoric of the benevolent superpower.

In his book, *Rethinking the Cold War,* journalist Eric Black writes:

> The 1950s American kid's Cold War paradigm went something like this: The Russians were atheistic, totalitarian Communists who wanted to conquer the world so they could take away everyone's freedoms. ...They were so powerful, relentless and ruthless that

> they might get away with it if we weren't careful. The good news was that the United States was even more powerful. We were Number One and the whole world loved us because all we wanted was peace, prosperity, freedom and democracy for everyone.

Writing before the breakup of the Soviet Union and the crash of the Berlin Wall, Black noted that people need paradigms in order to "attach meaning to events." He continued his assessment:

> Our paradigm is constantly reinforced by television shows, movies and plays, even by the vocabulary our leaders and our news media employ when they talk about the world. The United States has allies; the Soviet Union has puppets. Our side is run by governments, theirs by regimes. We have police, they have secret police. We engage in covert actions, they commit subversion. Our government puts out announcements, theirs propaganda.[69]

Black is right, it seems to me, in his assessment that people cling to paradigms. "Eventually," he says, "a framework that was supposed to help us understand reality distorts our perception of reality instead."[70] A rigid paradigm, in other words, functions as a kind of internal thought police. It tells us to ignore the evidence whenever our nation's conduct conflicts with the lofty ideals many of us have learned and deeply internalized — that is, whenever policies and practices don't fit the framework of the United States as a benevolent superpower working valiantly to oppose all the forces of evil.

Eric Black's analysis raises some important issues which are relevant for our probe of why the SOA does what it does.

> During the 1970s and 80s, the United States has supported governments in Latin America and elsewhere that rule by force, flout their own constitutions, rig elections, murder, torture and oppress their own people. The State Department calls these countries "emerging democracies," [and] praises their improving human rights record.... The superpower that, according to our view of ourselves, respects international law, defied international law by mining the harbors of [Nicaragua], then refused to defend its actions before the World Court of Justice. The CIA created, armed and trained the Contras... to overturn the Nicaraguan revolution.... Our ideals and our conduct in the world are harder and harder to reconcile. Our euphemisms and self-deceptions are harder to accept. And yet... the paradigm is alive and well.[71]

Is the School of the Americas' grisly association with numerous graduates who are guilty of human rights abuses an aberration that is rightfully filed away by our internal police? Or is it a piece of important evidence which reveals part of a consistent pattern within U.S. foreign policy, a pattern that we fail to see because, for whatever reason, we prefer to cling to rather than reassess our paradigm?

Tom Tomorrow, in a cartoon which appeared in the *Des Moines Sunday Register,*[72] powerfully and profoundly touches on issues central to the question of how and why the U.S. School of the Americas has trained and collaborated with torturers for more than three decades and why it has gotten away with it. The cartoon has four frames, each with a picture and two sets of words, one set describing an issue, another in which the figure or person portrayed speaks. The first frame offers the following description:

> The Pentagon runs a school in Fort Benning, Georgia, which has trained, among others, an organizer of Salvadoran *death squads,* the head of an Argentine *junta,* and *Manuel Noriega.*

Beneath it is a picture of a teacher standing in a classroom. In the background is a blackboard lined with letters from the alphabet. "Good morning, class!" the teacher says. "Today we'll be studying *advanced interrogation techniques!* Can *you* say cattle prod?"

The second frame continues to paint portraits of SOA graduates. It says:

> Another graduate is Col. *Julio Alpirez,* the Guatemalan officer on the CIA payroll who murdered an American hotelier and a rebel leader married to an American.... When Col. Alpirez's exploits were revealed by House Intelligence Committee member Robert Torricelli, *Newt Gingrich* responded *swiftly*... demanding the punishment of *Torricelli.*

The companion picture to this scene shows Gingrich at a press conference: "It's an *outrage*...," Gingrich says, "that this information was made public, I mean."

The third frame in Tom Tomorrow's cartoon describes damage control now that the bad news is out. It says:

> Now that the cat *is* out of the bag, a predictable scenario will unfold.... Hearings will be held... shock will be professed... and finally, the CIA will loudly and publicly declare the entire matter a *dreadful aberration* which was entirely the fault of a convenient, to-be-determined *scapegoat.*

Two men are shown below these words looking somewhat perplexed. The first says, "Um — you see — *Aldrich Ames* did it!" "Yeah," the other chimes in, "that's the ticket! That *scoundrel!*"

In the fourth and final frame of the cartoon a penguin stands, wearing hat and scarf, holding an umbrella, with a city sitting in the background. He says:

> The media will undoubtedly cooperate eagerly with this ritual of absolution... doing their best to ignore our country's ongoing complicity in the deaths of some *110,000 Guatemalans* at the hands of successive U.S.-backed dictatorships over the past 30 years.

The cartoon ends with the following commentary which closes out the fourth frame. "And if you didn't already know about that — well, at the risk of repeating ourselves — isn't it sad that you had to learn about it from a *talking penguin in a comic strip?*" (emphasis in original). Indeed!

— ❖ —

Let me briefly summarize what we have learned while probing a part of our country's hidden history. We discovered U.S. leaders and foreign policy planners building military alliances and fostering national security states in pursuit of stability. Democracy, human rights, and raising living standards, they said, were "unrealistic objectives," luxuries of "altruism and world-benefaction" we could not afford. They spoke about "envy and resentment" rooted in the vast gap separating the rich and poor and then assigned U.S. foreign policy the task of devising "a pattern of relationships which will permit us to maintain this... disparity." A powerful business leader described the U.S. as "the majority stockholder in this corporation known as the world," told his colleagues that "American private enterprise" would either "strike out and save its position all over the world, or sit by and witness its own funeral," and argued for a "foreign policy...concerned with the safety and stability of our foreign investments." We heard about an "implacable enemy" and the need for U.S. foreign policy to fight that enemy using any means necessary. There were "no rules in such a game," we were told, and no "accepted norms of human conduct" to follow. U.S. foreign policy needed "to subvert, sabotage and destroy our enemies by more clever, sophisticated, more effective methods than those used against us."

It would be nice if we could dismiss the above perspectives as the musings of a lunatic fringe. But we cannot. What they reveal are the philosophical, ideological, and practical foundations of U.S. foreign policy as articulated by those who used their power to shape that policy. Their perspectives, unfortunately, make the linkage between the School of the Americas, its graduates, and human rights atrocities entirely logical and understandable. What policies realistically emerge from the foreign policy perspectives articulated above? What role for the School of the Americas? What kind of SOA curricula? Lacking a commitment to democracy, human rights, or authentic development, U.S. foreign policy "by any means necessary" became an escalating spiral of violence. "You can't squeeze blood from a turnip," the old saying goes, nor democracy from the School of the Americas. You can, however, realistically expect to find real blood flowing freely from the river of terror, torture, and human rights violations that are the logical consequences of the policy perspectives discussed above. Not surprisingly, SOA curriculum itself reflects the "by any means necessary" philosophy.

Due to the faithful witness of SOA Watch protesters and Diana Ortiz we are beginning to get a better profile of the training manuals used at the School of the Americas. In response to Diana Ortiz's fast in front of the White House the Clinton administration ordered the Intelligence Oversight Board to investigate U.S. operations in Guatemala. Buried in the report was the following paragraph:

> Congress was also notified of the 1991 discovery by DOD [Department of Defense] that the School of the Americas and Southern Command had used improper instruction materials in training Latin American officers, including Guatemalans, from 1982 to 1991. These materials had never received proper DOD review, and certain passages appeared to condone (or could have been interpreted to condone) practices such as executions of guerrillas, extortion, physical abuse, coercion, and false imprisonment.[73]

Despite the language of damage control ("These materials had never received proper DOD review") the cat was now out of the bag. The Pentagon was eventually forced by public pressure to declassify the above mentioned manuals. The manuals advocate executions, torture, false arrest, blackmail, censorship, payment of bounty for murderers, and other forms of physical abuse against enemies. The manual on "Terrorism and the Urban Guerrilla," for example, says that "another function of the CI [counterintelligence] agents is recommending CI targets for neutralizing," a euphemism for elimination or assassination. Military officers were also taught to gag, bind, and blindfold suspects. These skills take on a more ominous meaning when we realize that they are part of the "Interrogation" manual and that many thousands of Latin Americans were tortured and murdered during interrogation while gagged, bound, and blindfolded. Another manual, "Handling of Sources," notes how the "CI agent could cause the arrest or detention of the employee's [informant's] parents, imprison the employee or give him a beating as part of the placement plan."[74]

The Pentagon acknowledged that the materials used in the manuals have a long history. Many of the training instructions were used in the 1960s by the Army's Foreign Intelligence Assistance Program, "Project X." Significantly, these manuals which advocate and teach torture and terror were widely distributed to countries in Central and South America where the U.S. was most heavily involved in counterinsurgency programs, including Guatemala, El Salvador, Honduras, and Panama.[75]

Representative Joseph Kennedy summarized the significance of the manuals. "The Pentagon," he said, "revealed what activists opposed to the school have been alleging for years — that foreign military officers were taught to torture and murder to achieve their political objectives."[76]

SOA supporters falsify the historical record of abuse traceable to SOA curricula. Even recent additions which touch on human rights and democracy are suspect. Retired U.S. Army Major Joseph Blair, cited earlier, joined more than four hundred protesters at Fort Benning in November 1966 to demand that the SOA be closed. He said that the school "is teaching everything that it has always taught plus a couple of courses which have been added to appease Congress and the press."[77]

The hidden history discussed above is part of a gruesome story shrouded in secrecy and eating at our national soul. There is, unfortunately, more bad news to tell. You have perhaps noticed a disconcerting pattern in the extensive evidence linking the School of the Americas and its graduates with human rights abuses, namely, that the church seems to be a specific target for repression. It is to this issue that we now turn.

Chapter 5

The Death of God

It is a crime to be a Christian and to demand justice.
—SALVADORAN DELEGATE OF THE WORD, APRIL 1988[78]

In previous chapters we have touched on disturbing connections between U.S. foreign policy, the School of the Americas, and religious persecution in El Salvador: SOA graduates played key roles in the murders of Archbishop Romero, the four U.S. church women, and the Jesuit priests; SOA graduate General Hector Gramajo was responsible for the abduction, rape and torture of Sister Diana Ortiz in Guatemala; and SOA Hall-of-Fame member General Hugo Banzer was known throughout Latin America for violence directed against progressive church leaders.

We also saw that for more than thirty years following World War II the United States secured its interests and sought stability by supporting military regimes throughout Latin America. The United States offered logistical, financial, military and ideological backing to national security states. Such states expected and required church support for their unequal and repressive regimes and they threatened and carried out bloody sanctions for those who violated their circumscribed role.

We heard former SOA instructor Retired U.S. Army Major Joseph Blair describe Latin American soldiers and trainers at the school expressing hostility for segments of the Catholic church which they associated pejoratively with human rights. Most Latin American soldiers, he said, equate human rights with subversives, particularly the Catholic church.[79] Even mock human rights exercises at the School of the Americas revealed hostility toward the church, as the priest and catechists get killed or roughed up most of the time. As funny as these results apparently seem to those at the School of the Americas, persecution of the churches is no joke.

America's Watch describes how the intensive repression against religious workers and organizations in El Salvador immediately following the Jesuit murders reflects a *consistent pattern of religious persecution throughout the war,* a war in which SOA graduates were responsible for numerous atrocities, including the murder of El Salvador's key religious leaders:

> The government's hostility towards church and relief organizations was particularly pronounced: In the period November 13–December 14 [1989], there were 54 searches of 40 different church facilities and homes of church workers by Salvadoran military and security forces. Dozens of church workers received death threats and fled the country under government order or death threat, dozens more . . . were jailed and abused in detention, and numerous church facilities were ransacked. . . . The symbolic significance of the Army's murder of the country's leading academic and religious figures cannot be overstated: the deaths signal that, once again, no one is safe from Army and death squad violence. . . . The Bush Administration has taken the position that the Jesuit murders were a dramatic departure from Salvadoran army policy. . . . In our view, the murders were en-

> tirely in keeping with Salvador's ten-year civil war.... Those responsible for almost every other instance of egregious abuse against Salvadoran citizens still enjoy absolute immunity.[80]

The substantial links between SOA graduates and persecution of progressive religious point toward a policy lurking beneath the shadows of secrecy. This and the following chapter will focus once again on the "why" question. *Why have U.S. foreign policy, the School of the Americas, and SOA graduates been so visibly involved in the murder and repression of church workers?* In an effort to answer this question we must once again look to history. In this case, I want to examine what I consider to be four watershed moments that help explain how religious persecution became central to U.S. policy and routinely carried out by SOA graduates: the Cuban revolution; changes in the Catholic church; the Nicaraguan revolution; and lessons learned from the U.S. defeat in Vietnam.

The first three of these snapshots or watershed moments will be developed in this chapter with the fourth being left for the chapter which follows. Each, it should be noted, could be the subject of lengthy discussion. I have provided only a bare-bones assessment, just enough, I hope, to shed light on how and why a U.S. policy of targeted repression against the church emerged and took its deadly toll. Understanding these four key snapshots in our recent historical experience will not make U.S. involvement and support for church repression less despicable. It will, however, make the policy behind the repression visible and comprehensible and, I hope, strengthen the case to close the School of the Americas.

The first watershed moment was the Cuban revolution of 1959. The overthrow of a U.S.-backed dictator and the establishment of a communist government in the hemisphere, not surprisingly, had far-reaching consequences both for U.S. foreign policy and Latin American governments.

The United States responded to the Cuban revolution with the Alliance for Progress, a program with two conflicting and, as it turned out, irreconcilable impulses and components. On the one hand, promoters of the Alliance recognized that poverty was a breeding ground for social turmoil. This led to a slight increase in economic aid and a large increase in reform rhetoric. On the other hand, the Alliance dramatically increased support for Latin American militaries and the focus of aid and training shifted further to warfare against internal enemies.

Whatever expectations of economic reform and development the Alliance for Progress may have generated were quickly dashed. Preoccupation with stability and fear of internal enemies translated into increased U.S. support for deadly counterinsurgency programs and repressive militaries. Senator Edward Kennedy summarized the disastrous results:

> Their [Latin American countries'] economic growth per capita is less than before the Alliance for Progress began; in the previous eight years U.S. business has repatriated $8.3 billion in private profits, more than three times the total of new investments; the land remains in the hands of a few; one-third of the rural labor force is unemployed and *13 constitutional governments have been overthrown since the Alliance was launched* (emphasis added).[81]

Castro's victory and the triumph of communism in Cuba also served to legitimize and strengthen the ideology and institutions of national security states. The fear of communism became the force driving U.S policy, shaping the School of the Americas, and distorting Latin American politics. An editorial in the *Atlanta Constitution* calling for closing the School of the Americas describes this dynamic while offering helpful clues into the school's abysmal human rights record:

> From its [SOA's] inception in 1946, some 57,000 soldiers from Latin America and the Caribbean have received instruction in counterinsurgency and commando tactics, military intelligence, psychological warfare, etc. The school's purpose initially was to promote internal stability throughout the region, but gradually it became focused on helping like-minded governments combat Communist subversion.[82]

The Cuban revolution also had an indirect yet profound impact on the church because it prompted priests, nuns, and lay workers from around the world to go to Latin America to join the fight against communism. What happened, however, is that many who came to serve in the ideological battle against atheistic communism found themselves working with poor people crushed by poverty and the distorted power and priorities of national security states. Many religious working with the poor came to see communism as an exaggerated threat that obscured the real causes of hunger and misery, namely, the unhealthy alliances between U.S. businesses, U.S.-backed militaries, and local economic elites. In short, many of those who came to fight communism discovered different enemies: hunger, disease, and the national security states responsible for political repression.

The second watershed moment had to do with profound changes in the Catholic church itself. Pope John XXIII had convened a meeting of bishops from around the world in

Rome in 1962 in order to reassess the role of the Catholic church. The meeting of the Second Vatican Council, or Vatican II, lasted for three years. It ushered the church belatedly into the twentieth century as it tried to find its proper role in the modern world.

More important, however, for understanding how a church so long identified with the privileged and powerful became the object of intense persecution was a meeting of Latin American bishops in Medellín, Colombia, in 1968. For centuries, with a number of notable exceptions of course, the Catholic church had aligned itself with the politically and economically powerful. The bishops at Medellín, however, were influenced by priests and nuns whose faith was shaped while working among the poor. The central question of Vatican II was how to make God and the church relevant to the modern world, the world of technology and progress. The bishops at Medellín confronted a different world. Latin American majorities lived in a world of misery, not progress.

The bishops who gathered at Medellín asked questions about God and the proper role of the church in a situation of death, for that is what poverty means to the poor. They called Latin America's unequal societies sinful, and cited concentrated land ownership and the vast gulf separating the rich and poor as examples of institutionalized violence which led to hunger and misery. The church's task in the midst of such sin, the bishops said, was to make a preferential option for the poor and to call the rich to conversion in an effort to free their societies from the bondage of sinful social structures.

Those seeking a church relevant to the poor and rooted in Latin American reality grounded their faith and struggle in a "theology of liberation." From the Exodus to the life and practice of Jesus, the God of the Bible is made known through efforts to free human beings from all that binds them, including the economic chains so visible in ancient Israel and throughout Latin America.

Liberation theology resonated in the experiences of the poor who had heard for too long that their poverty was God's will and that their ticket into heaven was passive acceptance of their misery. For others, however, liberation theology was subversive. Rich Latin Americans, like the vast majority of wealthy Christians everywhere, saw wealth as a sign of God's blessing. They found the association between wealth and sin highly offensive. Many members of the church hierarchy were also upset. They were afraid that "a church of the poor" might undermine their own hierarchical authority and might disturb comfortable links between the church and the social and economic elite.

The military, of course, despised liberation theology and the progressive churches. National security states, after all, were held together in part because the church had willingly joined the military, the oligarchy, and big business in a cozy alliance of mutual appreciation. The generals hated the theologians who articulated liberation theology, the priests, nuns, and lay workers who lived it out in the midst of the people, and the people themselves who, inspired by faith in God and each other, organized on behalf of a better future.

Penny Lernoux describes how organized attacks against progressive religious escalated throughout Latin America a few years following Medellín. At the center of these attacks was SOA graduate and Hall-of-Fame member Hugo Banzer:

> Archbishop Alfonso López Trujillo...denounced a "concerted campaign against the Church." While

there were plenty of skirmishes between national churches and governments in Latin America during the early 1970s, only since 1975 has there been a marked similarity in these anti-Church campaigns, such as planting communist literature on Church premises or arresting foreign priests and bishops on trumped-up charges of "subversion." This was no coincidence: at least eleven countries were following the same geopolitical plan, the Doctrine of National Security, and therefore shared similar strategies in their "war on communism."

A typical contribution to this common strategy was the "Banzer Plan," hatched in the Bolivian Interior Ministry in early 1975 and named for Hugo Banzer, Bolivia's right-wing military dictator. Although the original plan was not committed to paper, it was discussed at length in the Interior Ministry, a publicly acknowledged subsidiary of the CIA, after the Bolivian Church began to make trouble for the government by denouncing the massacre of tin miners. The plan was leaked to Bolivia's Jesuits by an Interior Ministry official, who was horrified by the government's intention to smear, arrest, expel, or murder any dissident priest or bishop in the Bolivian Church. The authenticity of the Banzer Plan, which boasted many of the classic "dirty tricks" employed by the CIA in Latin America during the 1960s, was subsequently confirmed by the government itself when it followed, word for word, all the original tactics. . . . The three main thrusts of the campaign were to sharpen internal divisions within the Church, to smear and harass progressive Bolivian Church leaders, and to arrest or expel foreign priests and nuns, who make up 85 percent of the Bolivian clergy.[83]

U.S. foreign policy planners, not surprisingly, joined and in many cases led the chorus of alarm concerning changes in the church. Out of touch with the complex social winds blowing around them, U.S. foreign policy planners fell back on ideology, seeing "Marxist-Leninist forces" utilizing "the church as a political weapon against private property and productive capitalism. . . ."[84] One of the Pentagon training manuals, which instructed Latin American officers in the rationale and techniques of torture and execution, states: "The CI [counterintelligence] agent must consider all the organizations as possible guerilla sympathizers."[85]

In sum, liberation theology and a changing church provide essential background as to why the School of the Americas, SOA graduates, and U.S. foreign policy generally targeted the church for repression and why many SOA graduates have been enthusiastic students of terror.

The third watershed moment was the Nicaraguan revolution. In 1979, a U.S.-sponsored family dictatorship that had ruled Nicaragua for five decades was suddenly and unexpectedly overthrown by a popular revolution. For many poor people in Central America this was a reason to celebrate and a time of profound hope. A dictator and repressive military backed by the world's most powerful country had fallen and a revolutionary government committed to redistributing land and creating opportunities for the poor had come to power. At the same time, throughout Central America, a significant segment of the church had broken its deadly alliance with the rich and powerful to become a church of, with and among the poor, sharing common aspirations and, as it turned out, a common fate.

Central America during the decades of dictators and national security states was dominated by hunger and inequality. The United States had backed repressive militaries, unrepresentative governments, large landowners, and big business interests at the expense of popular aspirations and much needed social change. U.S. foreign policy might have been reassessed in light of the Nicaraguan revolution, but instead it fell victim to distorted ideology and exaggerated fears.

U.S. foreign policy planners and their Central American allies saw the hope of the poor and the emergence of a prophetic church as evidence of a communist conspiracy. As a result, two tiny impoverished countries, Nicaragua and El Salvador, became the most important places on earth for U.S. foreign policy planners who were determined to counter "communist subversion," whatever the financial and human cost.

According to official rhetoric of the day, Central America and the Caribbean were under attack by an international communist menace that threatened not only U.S. regional interests but our own society as well. The Council for Inter-American Security published a paper in 1980, known commonly as the Santa Fe Report, that was typical in its tone and content. It noted that the "young Caribbean republics situated in our strategic backyard face not only the natural growing pains of young nationhood but the dedicated, irrepressible activity of a Soviet-backed Cuba to win ultimately total hegemony over this region. And this region," the report continued, "is the 'soft underbelly of the United States.' "[86]

The School of the Americas became closely aligned with dictators and torturers in part because reprehensible policies sprout from the fertile ground of rhetorical excesses. The Santa Fe Report, for example, betrayed a lack of even the most basic knowledge of the forces driving the deep desire for social change throughout Central America. Typical of its simplistic distortions was its portrayal of liberation theology:

> U.S. foreign policy must begin to counter... liberation theology as it is utilized in Latin America by the "liberation theology" clergy.... Unfortunately, Marxist-Leninist forces have utilized the church as a political weapon against private property and productive capitalism by infiltrating the religious community with ideas that are less Christian than communist.[87]

The Santa Fe Report dismissed liberation theology as another manifestation of communist subversion. Crazy as this might seem, it was not the product of a lunatic fringe. The Santa Fe Report became the foreign policy blueprint for both the Reagan and Bush administrations. Given the strength of its distorted convictions we should not be surprised that ideologically led rhetoric translated into deadly policies and blood-stained hands that left a trail to and from the School of the Americas. The subversives "are who they've always been," Father Roy Bourgeois says. "Not communists, but religious and health-care workers." The School of the Americas says it wants to professionalize Latin American armies, but "the military is too damn professional now."[88]

It is more than coincidence that the Santa Fe Report was published in 1980, the same year Archbishop Romero and the four U.S. churchwomen were murdered by SOA graduates. It is more than coincidence, as America's Watch reports, that "the Jesuit murders" and widespread persecution of the churches "were entirely in keeping with Salvador's ten-year civil war." And, it is more than coincidence that Ignacio Ellacuría and the other Jesuits were murdered by SOA graduates after a 1987 report from the Conference of American Armies, a meeting which brought together commanders from the United States, Argentina, Uruguay, Paraguay, Bolivia, Brazil, Peru, Ecuador,

Colombia, Venezuela, Panama, Honduras, Guatemala and El Salvador, named liberation theology and the progressive churches as the principal security threats in the Americas. The conference report specifically targeted Father Ellacuría, naming him as one who consciously manipulated "the truly liberating Christian message of salvation to further the objectives of the Communist revolution."[89]

To fully understand the policy behind the persecution of the church, however, we need to examine our fourth watershed moment, the U.S. defeat in Vietnam and how a reassessment of that war led to the deadly strategy of low-intensity conflict.

Chapter 6

Religious Persecution and Low-Intensity Conflict

> It takes relatively few people and little support to disrupt the internal peace and economic stability of a small country.
>
> —William Casey, CIA Director[90]

Religious persecution *as policy* is even more understandable in the context of low-intensity conflict (LIC). To understand LIC, however, we must revisit the Vietnam war, our fourth watershed moment. Vietnam is important for our present probe because the U.S. defeat in Vietnam led U.S. policy makers to refine counterinsurgency techniques that were implemented with deadly efficiency throughout Central America in the 1980s following the Nicaraguan revolution.

U.S. involvement in Vietnam began clandestinely in the 1950s and ended in humiliating defeat in the 1970s. In *War against the Poor: Low-Intensity Conflict and Christian Faith* I describe six key lessons learned by foreign policy planners in light of the U.S. defeat in Vietnam:

Lesson 1: Improve military capacity. Fighting wars effectively in the Third World meant development of quick-strike or Special Operations Forces. In other words, U.S. military forces and training appropriate for fighting a nuclear or conventional war against the Soviet Union were ineffective in third world settings. Neil Livingstone, a Pentagon consultant on low-intensity conflict, said that in the post-Vietnam period the "security of the United States requires a restructuring of our war-making capabilities, placing new emphasis on the ability to fight a succession of limited wars, and to project power into the Third World."[91]

Lesson 2: Be cost effective and control hearts and minds. Huge amounts of money spent and an almost unlimited numbers of bombs dropped did not lead to victory in Vietnam. We lost the war, according to this view, because we did not either win over the hearts and minds of the people, or sufficiently crush their spirits.

Lesson 3: Let others do the dying. A key challenge facing U.S. policymakers in the post-Vietnam period was how to defend U.S. interests in the Third World while limiting U.S. casualties. Large numbers of U.S. soldiers returning in body bags during the Vietnam war resulted in public opposition. Michael Klare perceptively describes low-intensity conflict in this way:

> Low-intensity conflict [LIC], by definition, is that amount of murder, mutilation, torture, rape, and savagery that is sustainable without triggering widespread public disapproval at home. Or to put it another way, LIC is the ultimate in "yuppie" warfare — it allows privileged Americans to go on buying condominiums, wearing chic designer clothes, eating expensive meals at posh restaurants, and generally living in style without risking their own lives, without facing conscription, without paying higher taxes, and, most important, without being overly distracted by grisly scenes on the television set. That, essentially, is the determining characteristic of low-intensity conflict in the Americas today.[92]

Lesson 4: Repression and terror are sometimes useful but must be managed carefully. From the perspective of U.S. policy planners the focus of our training is to help our military and police force allies maintain stability. Repression and terror are neither good nor bad but they are sometimes necessary. Repression rises or falls in response to the strength of the popular movements. Accordingly, repression, including torturing political opponents, can be necessary in one setting and counterproductive in another.

Lesson 5: In warfare victory has many meanings. I said earlier that the U.S. defeat in Vietnam prompted a reassessment. One of the things reassessed was the definition of victory or defeat. On one level the U.S. lost the Vietnam war because Saigon fell, the U.S. was humiliated, and Vietnam was outside U.S. control. On another level, those who revisited Vietnam concluded that the U.S. on some levels had won the war. The war had effectively destroyed Vietnam's economy so that it might never recover. Any subsequent failure in Vietnam could be further evidence of the failures of socialism.

In Central America this lesson was applied to Nicaragua. As CIA Director William Casey said, "It takes relatively few people and little support to disrupt the internal peace and economic stability of a small country." Casey acknowledged that the U.S. war against Nicaragua might not succeed in overthrowing the revolutionary government but whatever the outcome we could "waste" the country.[93] The Executive Summary Report of a U.S. Medical Task Force investigating U.S.-supported contra attacks against civilians in Nicaragua noted: "It is abhorrent that a primary goal of the contra army is the systematic destruction of the Nicaraguan rural health care system."[94] America's Watch reported that contra violations of the laws of war were "so prevalent that these may be said to be their principal means of waging war."[95]

Lesson 6: Deceive your own people. The U.S. "lost" the war in Vietnam, according to the post-war assessment, because the liberal press brought disturbing images of the war into the living rooms of most Americans who found it increasingly difficult to reconcile official rhetoric with reality. This lesson about the importance of secrecy and deception was applied rigorously in Central America throughout the 1980s as part of low-intensity conflict strategy. Low-intensity conflict was meant to be low-visibility warfare. "What you have," Morton Halperin said during a time in which the Reagan administration was illegally funding the Nicaraguan contras, trading arms for hostages in the Middle East, secretly supporting Salvadoran death squads, training numerous soldiers at the School of the Americas, and facilitating the work of drug traffickers in exchange for foreign policy favors, "is a growing gap between the perceptions inside the executive branch about what the threats are to our national security, and the beliefs in the . . . public about the threats to national security."

Halperin, who once resigned his staff position on the National Security Council in protest over U.S. policy in Vietnam and Cambodia, writes:

> [The gap in perception over national security needs] leads to secrecy; that is what drives policy underground, that's what leads the president to rely more on covert operations, what leads the president and his officials to lie to the public, then lie to the Congress about the operation. Precisely because they cannot get their way in public debate, they are driven to circumvent the democratic process.[96]

These six lessons learned from the Vietnam war have profound implications for Central America and for our understanding of U.S foreign policy and the involvement of the School of the Americas and SOA graduates in religious persecution. Each lesson was incorporated into low-intensity conflict strategy in which four aspects of warfare are integrated into a flexible and comprehensive package. In a nutshell, LIC strategy says that warfare in third world settings such as El Salvador and Nicaragua must be fought simultaneously on four basic fronts.

The first front is economic. The United States, utilizing LIC strategy, uses economic leverage, including rewards and punishments, to pressure allies and adversaries. In El Salvador we propped up unpopular governments with massive economic aid and used economic leverage to force them to make structural reforms favorable to U.S. business interests. In the case of Honduras, the United States used the economic dependency and indebtedness of that country as leverage to gain access to Honduran territory as the training ground and safe-haven for the Nicaraguan contras fighting on our behalf to destroy the Nicaraguan revolution. In Nicaragua, in our effort to destroy a popular revolution, the United States blocked loans, slapped on an embargo, trained the contras in destabilization tactics, and forced the country to divert resources away from economic development to the military sector.

A second front within LIC strategy is military. The U.S. created, funded, trained, and equipped the Nicaraguan contras. Our own quick strike forces illegally mined Nicaragua's harbors. We trained numerous Salvadoran and other soldiers at the School of the Americas and elsewhere. We provided training as well as financial, logistical, and weapons support to the Salvadoran army.

A third front within LIC strategy is diplomatic. The

U.S. lobbied our European allies to reduce support to Nicaragua and to quiet their protests against U.S. policy in the region. Our government sought to discredit Nicaragua's elections while carrying out sham elections in El Salvador which served to cloak the real power residing in the military and within the U.S. Embassy. The U.S. blocked a negotiated settlement to regional conflicts for most of the decade of the 1980s. It sought a negotiated settlement to conflicts in the region only after Nicaragua had been economically destroyed and politically defeated.

Finally, a fourth front within LIC strategy is psychological. This fourth front requires more attention because it is the most relevant for our understanding of religious persecution as policy. The key component in psychological warfare is the use and management of terror. SOA supporters claim that El Salvador and Honduras are SOA success stories. The SOA's emphasis on human rights, according to this view, resulted in a significant reduction in the number of human rights violations by the end of the 1980s. This perspective is a blatant perversion of history. One lesson U.S. policymakers learned from Vietnam was that not all terror is useful and that terror can sometimes be counterproductive. SOA "human rights training" was in fact a compact with the devil in which the School of the Americas tried to help Latin American officers and soldiers learn the art of managed terror. In *War against the Poor* I used the following analogy to describe the management of terror and to illustrate the difference between managed terror and respect for human rights:

> Imagine a situation in which mass murderers kill people in your politically active neighborhood for eight consecutive weeks. Among the dead are both neighborhood activists and others less active but pos-

> sibly sympathetic to the ideas of the activists. Human rights groups within and outside your neighborhood protest against the violence. After eight weeks of generalized terror, daily funerals, and blood in the streets there is a significant reduction in the overt use of violence. "Only" five people are killed weekly during weeks nine and ten. All of the victims were apparently targeted for assassination because they were members of neighborhood organizations or members of local human rights groups that had demanded the perpetrators of the violence be brought to justice. The U.S. government cites reduced numbers of death-squad victims as "proof" of its commitment to human rights in El Salvador.... The following three questions, based on the analogy above, illustrate the difference between respect for human rights and the management of terror....
>
> Would a reduced body count make you and your family feel safe in your neighborhood if not one of the mass murderers had been arrested, tried before a court of law, or jailed...? Would a reduction in assassinations from thirty to five each week encourage you to be involved politically if you knew that while the body-count figures were down activists were being targeted...? What would be your response if several of your neighbors took advantage of the "safer conditions in the neighborhood" and spoke out freely, only to be killed (so that in subsequent weeks the numbers of dead averaged fifteen)?[97]

When I was in Central America in the mid-1980s it was widely believed that Vice-President George Bush and other U.S. officials had lectured El Salvador's death squads, not because they wanted to shut them down but because they differed over how much overt terror was needed at that particular moment during the war. For many years U.S. leaders had enthusiastically supported Salvadoran government and military leaders as they implemented policies of massive terror. At some point during the mid-1980s, however, U.S. leaders came to believe that massive terror was no longer needed and was in fact counterproductive. The death squads and many of El Salvador's key military leaders to whom they were connected, apparently weren't so sure.

I do not know whether George Bush actually met with the death squad leaders. I do know that this account of events is widely held and that it reflects the actual experiences of the people. In other words, the United States was deeply involved in managing the war, a war that included terror tactics utilized by SOA graduates against popular sectors, including the church. Several other examples should be sufficient to illustrate the central role of terrorism within LIC's emphasis on psychological warfare. The Central Intelligence Agency produced a manual for the Nicaraguan contras which included instructions on "Implicit and Explicit Terror." It encouraged the neutralization (assassination) of "government officials and sympathizers," just as other Pentagon training manuals used at the School of the Americas did. Edgar Chamorro, a former contra leader, testified before the World Court that "the practices advocated in the manual," which was titled *Psychological Operations in Guerrilla Warfare,* "were employed by the F.D.N. [contra] troops. Many civilians were killed in cold blood. Many others were tortured, mutilated, raped, robbed or otherwise abused."[98] Former CIA official John Stockwell said in relation to U.S. strategy in Nicaragua that "encouraging techniques of raping women and executing men and children is *a coordinated policy of the destabilization program*" (emphasis added).[99]

It is not my intent to offer a detailed look into low-intensity conflict strategy. Interested readers can find more detailed information elsewhere.[100] It is important, however, to know something about the evolution of LIC strategy and the importance of psychological warfare if we are to explain why U.S. foreign policy, the School of the Americas, and SOA graduates targeted the progressive churches for repression and terror.

Perhaps church repression can be better understood if we imagine the frustration, disappointment, or rage that many U.S. policymakers must have felt during the late 1970s and early 1980s. Their predecessors had warned of implacable enemies and set out to defeat them. The national security states and dictators expected to hold these enemies at bay and to maintain stability, however, were failing. Stability without justice had proven to be an elusive goal and social turmoil was brewing throughout Central and Latin America. Castro, despite numerous U.S. attempts to oust him, had successfully established himself in Cuba and both he and the revolution were of intense interest to non-elite sectors throughout Latin America. Fascination with Cuba seemed to grow the more the United States vilified the revolution. This was due in part to the fact that the United States, which many in Latin America regarded as a modern-day Roman empire, treated Cuba as forbidden fruit. This fascination was also linked to Cuba's relative success in feeding, educating, and meeting the medical needs of its people. Food, health care, and schools were rare commodities for poor majorities living within Latin America's repressive societies.

Many U.S. foreign policy planners, the bitter taste of defeat in Indochina still lingering in their mouths, had sought redemption in Central America. Their ample reasons for frustration multiplied when the revolutionary government which ousted a U.S.-backed dictator in Nicaragua was validated at the ballot box and when a revolutionary movement seemed poised to take power in El Salvador. Even the church had apparently forgotten its proper role and place. Events were spinning out of control.

A particular visit to the U.S. Embassy in El Salvador stands out in my mind as a perfect example of the rage and frustration felt by many U.S. policymakers and their elite allies. First, a little background. Embassy briefings were forums in which U.S. officials, sometimes ambassadors, usually political officers, presented visiting U.S. citizens with a summary of official U.S. government policy toward the particular country in which they served. Such briefings were off the record, which I understood to mean you could write about what you heard at the embassy but you could not attribute it to the specific individual from whom you heard it. This, we were told, fostered a more open exchange because embassy personnel didn't have to be as guarded. Embassy briefings were always informative, though sometimes painful. I always found it frustrating, for example, to hear glowing reports from the U.S. Embassy about the dramatically improving human rights situation in El Salvador, always attributed to our country's enlightened policy, when prior or subsequent meetings with the "Mothers of the Disappeared," Tutela Legal (the human rights group of the Catholic archdiocese), or other independent human rights monitors painted a decidedly different picture.

The particular meeting I have in mind was a briefing on the human rights situation with a political officer from the U.S. Embassy in El Salvador. He sensed, correctly I think, that we were not buying his rosy portrait of human rights improvements. This was apparently not his first experience with a skeptical audience. For whatever reason, however, on this particular occasion our otherwise composed diplomat lost it. He began screaming at our group. (I mark his words with quotation marks but they should be understood as my paraphrase based on notes from the time of the meeting.)

"Alright," he said, his voice rising and his cheeks turning red with rage, "what is it that troubles you most? You didn't like the massive repression of several years ago when dozens of bodies appeared every day in the street. Well, if you didn't like massive terror," he continued, "remember this. WE-HAD-NO-OTHER-CHOICE!" Each word was spoken slowly, clearly, and loudly like a parent reprimanding a naughty child. "Nicaragua," he continued, "had already fallen and a massively popular revolution was about to take power in El Salvador. We had no other choice," he repeated. Then he stopped and regained his composure. A disturbing moment of truth-telling had passed. His expressed rage had swirled around the room like an uninvited demon shedding light on a dirty secret.

Perhaps most disturbing, from the vantage point of U.S. foreign policy planners, is that all of the setbacks and defeats named above occurred within the ideological backdrop presented earlier. What do you do when there have been no strings to bind your hands and you are still losing ground? The simple answer is you escalate the violence through low-intensity conflict.

In seeking to understand the relationship between U.S. foreign policy, the School of the Americas, and persecution of progressive religious we also need to remember that the SOA is a combat school which emphasizes counterinsurgency training. Counterinsurgency is warfare against *internal enemies.* Many Latin American officers teaching at the school, their U.S. counterparts, and their students include the church among those internal enemies, as does national security state ideology itself. "The persecution of the church," Archbishop Romero said, "is a result of defending the poor. Our persecution is nothing more nor less than sharing in the destiny of the poor."[101]

The progressive churches, their leaders, and participants were not only legitimate targets; they became, according to the Conference of American Armies, mentioned in the previous chapter, *the main target.* No wonder so many SOA graduates have watered the soil with the blood of so many priests and prophets. No wonder also that in mock exercises at the School of the Americas today, the priest and catechists often get killed.

In his book, *Powderburns: Cocaine, Contras and the Drug War,* former Drug Enforcement Administration agent Celerino Castillo III offers compelling evidence that death squads and religious persecution were products of official U.S. policy:

> Lt. Col. Alberto Adame, a U.S. military advisor to El Salvador . . . recommended one of his friends as a firearms instructor. . . . Dr. Hector Antonio Regalado, a San Salvador dentist, was a household name in the country's power corridors. I was shaking hands with "Dr. Death," as he was known in U.S. political circles, the man reputed to be the Salvadoran death squads' most feared interrogator. In El Salvador, he was known simply as "El Doctor." Regalado's prestige among the right wing stemmed from his ability to extract teeth — and information — without anesthesia. I wanted no part of *El Doctor.* I asked Adame if the embassy had approved Regalado as an advisor. He said Col. James Steele, the U.S. Military Group commander in El Salvador, gave Regalado his blessing. The military obviously wanted this man aboard, human rights abuses and all. . . . *El Doctor* harbored a boiling hatred for anything associated with Commu-

nism or revolutionaries, and showed particular disdain for the clergy, who sympathized with the peasants.

Castillo offers this description of the cynical practices of "Dr. Death" whose superior was SOA graduate Roberto D'Aubisson:

> Regalado painted a vivid picture of the death squads' modus operandi. After watching their intended victims for a few days to learn their movements, a dozen men in two vans would move in for the abduction. They preferred to strike away from the victim's home, bolting through sliding doors on both sides of the van and yanking the person off the street. As the torture began, they wrote down every name their victim cried out. Regalado practiced his impromptu dentistry on the unfortunate captives with a pair of pliers. I could see these doomed, bleeding men, screaming names with faint hope their pain would end if they fed their captors enough future victims. The pain usually ended with a bullet or the edge of a blade.... Regalado was convinced the clergy were Communist infiltrators, trained in Cuba to undermine El Salvador. He considered them cowards, hiding behind the cloth as they spread their diseased doctrine to the peasants he loathed. He spoke of personally directing the deaths of several outspoken priests.[102]

It seems appropriate to close this chapter with words from Archbishop Romero, whose murder is traced to SOA graduates and whose self-confessed killer, according to Castillo, was Dr. Death himself.[103]

> I want to make a special appeal to soldiers, national guardsmen, and policemen: Brothers, each one of you is one of us. We are the same people. The *campesinos*

> you kill are your own brothers and sisters. When you hear the words of a man telling you to kill, remember instead the words of God: "Thou shalt not kill." No soldier is obliged to obey an order contrary to the law of God.... In the name of God, in the name of our tormented people who have suffered so much and whose laments cry out to heaven, I beseech you, I beg you, I order you in the name of God, stop the repression![104]

If he were alive today I have little doubt that Romero would add: *Close the School of the Americas!*

Chapter 7

Damage Control

> Well, I'll be dipped in snuff, I said. I slapped my forehead in amazement. The cause of my wonder was a *Washington Post*...report that the U.S. Department of Defense admitted training Latin American military leaders in the arts of torture, execution, blackmail and other forms of coercion....Let me be more precise: The source of my amazement was not learning that my government has used my tax money to train Latin American thugs in torture, execution and blackmail....What astonished me is that it was in the news. In the *New York Times.* On "Dateline." The subject of indignant editorials....If you want a shining example of what's wrong with the American media, try this tale on for size. The open sewer called the School of the Americas at Fort Benning, Georgia, has been known to anyone who cared to find out about it for years. And years and years....And your news media are then shocked, *shocked,* to learn all this.
>
> —Molly Ivins[105]

The record of abuse traceable to the U.S. Army School of the Americas is so saturated with blood that it has wetted the pens of editorial boards throughout the country. The *San Antonio Express-News* calls the SOA a "breeding ground for human rights abusers."[106] The *Des Moines Register* says it is an "ugly thing" that the SOA serves to "advance the education of killers."[107] According to the *Atlanta Constitution* the decision to close the SOA "should be an easy one" because it "has strung together such a perverse 'honor roll' of cold-blooded murderers that America's meanest prison might be pressed to match it."[108] The *New York Times* adds its weight to this chorus of condemnation when it urges President Clinton, who supports the School of the Americas, to close it because doing so would "make it clear that in the future rogue operators, who abuse their relationship with the United States, will be exposed rather than protected." Closing the school, the *Times* continues, would "announce that America will no longer train and encourage Latin American thugs."[109] "SOA's best known products," the *Cleveland Plain Dealer* says, "have shared a distressing tendency to show up as dictators or as leaders or members of death squads. They have been agents of oppression."[110]

The Pentagon and CIA are not accustomed to such harsh words from the mainstream media which, as Molly Ivins implies above, is usually a willing accomplice in the nation's historical amnesia. After the Pentagon was forced to release the content of its training manuals U.S. military leaders and SOA supporters predictably fell back on damage control. Just as the CIA manual encouraging the Nicaraguan contras to assassinate political opponents and use torture was dismissed as an unfortunate mistake, the Pentagon training manuals would be explained away with similar arguments. The President's Intelligence Oversight Board said the "materials had never received proper DOD [Department of Defense] review." They only "appeared to condone" or "could have been interpreted to condone" the "executions of guerrillas, extortion, physical abuse, coercion, and false imprisonment." Once the unfortunate "error" had been discovered the DOD quickly "replaced and modified the materials."[111] "The problem was discovered in 1992, properly reported and fixed," said Lt. Col. Arne Owens, a Pentagon spokesman. A master of damage control, Owens sounded almost cheery. "There have been a lot of great changes at the School of the Americas."[112]

There are two important issues raised by the Pentagon's efforts at damage control and the media's sudden recourse to honesty. The first is the question of what we are to make of the unusual stream of editorial honesty. Why, in other words, at this particular time, has the mainstream press told the truth about the School of the Americas and called for its closure? I will return to this question after raising and dealing with the second issue, namely, the problem of historical amnesia itself.

Only ideological blinders or willful ignorance can allow us to dismiss as exceptional the huge body of evidence linking U.S. foreign policy, the School of the Americas, and SOA graduates to human rights atrocities. Effusive rhetoric crumbles beneath the weight of actual practices. Torture, terror, assassinations, persecution of the churches, and repression can't be dismissed as unfortunate mistakes or exceptional behaviors. They are, as I demonstrated in previous chapters, the fruits of calculated policy.

One reason damage control can no longer work is because there is too much damage and too much evidence of U.S. control. In light of the evidence, U.S. protestations over human rights abuses must appear to our Latin American allies as hypocritical, arrogant and self-serving. The United States declared their countries a battleground with the Soviets. It trained their militaries to be the guardians of our treasured stability. It shaped and gave legitimacy to national security state ideology and anti-communist hysteria. U.S. military and foreign policy planners descended on

Central America seeking redemption from Vietnam. They came to purge themselves of demons, to rescue their careers, and to test new theories of warfare. They defined liberation theology and the progressive churches as vile enemies to be defeated using any means necessary. They used training manuals that advocate terror and torture and their students implemented the lesson plans with cynical precision. Our leaders collaborated with and served as cover for death squads, and then expressed concern when disputes arose over how much terror was required at any given political moment.

All this was and is done in the name of freedom and democracy. Unwilling to face the truth about U.S. foreign policy and SOA complicity with terrorists, SOA supporters lifted up El Salvador as an SOA success story. SOA defenders boast about presidential and other distinguished alumni, relegating the number of SOA graduates who seized power unconstitutionally to the realm of unwritten footnotes. The self-serving mythology of the benevolent superpower lives on no matter how many torturers learn their craft at the SOA, how many religious leaders die at the hands of SOA graduates, how many dictators are embraced, how many blood-soaked bodies stain Latin American soil. This rhetoric, if merely tactical, is still dangerous and deadly. In the event that it is heart-felt it indicates a sickness in the depths of our national soul. Although an honest appraisal of U.S. foreign policy and the School of the Americas is painful, it is less costly than the self-deception which threatens democracy at home and abroad.

We can now return to the question of why major U.S. newspapers are telling a story that "for years and years" they refused to tell. As columnist Molly Ivins noted above, what is shocking to her as a journalist is not that the School of the Americas trains torturers but that the mainstream media is finally and belatedly telling us about it.

There are three important and related factors that help explain the mainstream media's turnaround. First, determined efforts by SOA Watch, other organizers, and grassroots supporters of the campaign to close the SOA have forced damaging information, long secret or underreported, into the public arena and consciousness. It has been one of the goals and one of the significant achievements of SOA Watch to get newspapers to take a stand on their editorial pages in favor of closing the SOA. When the School of the Americas is finally shut down we will owe an enormous debt of gratitude to these editorialists and to all who encouraged them to speak out.

Second, the sheer weight of the evidence implicating the School of the Americas and SOA graduates with atrocities is so overwhelming that it can't be ignored. Even before the Pentagon was forced to divulge the content of its training manuals — the "smoking gun" which proves SOA training of torturers and murderers — there were ample reasons for closing the school. Revelations found in the manuals, however, prompted a new wave of editorials. The *New York Times* is representative:

> Americans can now read for themselves some of the noxious lessons the United States Army taught to thousands of Latin American military and police officers at the School of the Americas....A training manual recently released by the Pentagon recommended interrogation techniques like torture, execution, blackmail and arresting the relatives of those being questioned....
>
> The newly released manual recalls a training manual that the Central Intelligence Agency distributed

> to the Nicaraguan contras...that recommended kidnappings, assassinations, blackmail and the hiring of professional criminals. The Reagan Administration quickly disowned that booklet when its contents were disclosed. Yet the School of the Americas continued to advocate similar methods for another decade. An institution so clearly out of tune with American values and so stubbornly immune to reform should be shut down without further delay.[113]

In sum, the flurry of editorial honesty concerning the School of the Americas can be traced in part to the faithful public witness of protesters and the overwhelming body of evidence linking the school, its training manuals, and its graduates to abuses.

A third explanation for the stream of editorials critical of the SOA is that the School of the Americas is *no longer useful.* It is possible, in other words, to lament past abuses and call for the school's closing because *at this particular moment in history the foundations which ground U.S. foreign policy have shifted!*

The editorial in the *New York Times* cited above states that "the school *does little to advance American interests* and should be closed down" (emphasis added).[114] Another SOA critic writes that Latin American countries "need trained police, not armies with skilled snipers and masters of psychological warfare. This is especially true *with the demise of any communist threat in Latin America*" (emphasis added).[115] "The school could maintain a veneer of importance *during the Cold War,*" an editorial in the *Bangor Daily News* says, "when communists were said to be marching toward the U.S. southern border. *Now the school is as relevant as the Berlin Wall...*" (emphasis added).[116] A *Washington Post* editorial condemning the CIA's coverup of recent atrocities in Guatemala committed by SOA graduate and CIA asset Julio Alpirez is even more direct. It says that it "defies credulity that, *at this late date in the United States's Central American involvement,* the CIA could still be recruiting killers of the sort that have made Guatemala's the region's bloodiest army" (emphasis added).[117]

The argument that the School of the Americas is *no longer useful* makes its way into many editorials. For some writers, its inclusion is part of an effort to broaden the appeal of critics who want to close the school. For many others, however, it is more, and there are two vital issues at stake.

First, a school that is no longer useful may very well have been useful *in the past.* Molly Ivins, in the quote at the beginning of this chapter, expresses shock and dismay because the mainstream media that is now boldly condemning the School of the Americas remained silent for so long. The mainstream media, including most of the newspapers which have recently editorialized against the School of the Americas, offered general and, in many cases, uncritical support for U.S. foreign policy objectives throughout the Cold War period. Our historical amnesia is rooted in their coverage. They reinforced at every turn the anticommunist hysteria which dominated U.S. foreign policy and gave impetus to the disastrous practices that have become the hallmark of the School of the Americas. In other words, the mainstream media that supported U.S. foreign policy throughout the Cold War period cannot divorce itself from the heinous deeds of the School of the Americas and its graduates. The media now seeks to discredit and close a school that is the product of policies it consistently defended!

The second critical issue can be posed as a question: If the School of the Americas was useful in the past, then why do many of its previous supporters feel it isn't useful now? Why so much present indignation from the mainstream press after "years and years" of deadly silence? Or stated more simply, What has changed?

In seeking to answer these questions we will need to return briefly to our discussion of low-intensity conflict strategy and to the tactical differences that surfaced in El Salvador concerning the management of terror. An enraged U.S. Embassy spokesperson, you will recall, defended a period of massive terror in El Salvador because a revolutionary government had ousted a U.S.-backed dictator in Nicaragua and a revolution was about to succeed in El Salvador. The United States, he said, "had no other choice." Later in the war, when U.S. officials felt they had more choices, they apparently coerced reluctant Salvadoran death-squad leaders and their supporters in the Salvadoran military to shift from massive terror to more selective violence.

— ❖ —

Recall also that low-intensity conflict strategy integrates economic, military, political-diplomatic, and psychological aspects of warfare into a comprehensive whole. LIC, in other words, offers a smorgasbord of options which are far-reaching and flexible. Military aspects of LIC predominated during some periods of the war. At other times, psychological warfare, including terror and torture, reigned supreme. Toward the end of the 1980s, however, economic forms of control had risen to the pinnacle of U.S. power in its relations with El Salvador and nations throughout the Third World. Economic leverage gave the United States more choices. It enabled the United States to put its weight behind a negotiated settlement in El Salvador and it enables the United States to exercise power throughout the Third World.

The School of the Americas is now considered expendable by many former supporters for the same reason that the United States today prefers elections in third world countries to dictatorial rule: economic leverage gives the United States sufficient power to defend its "vital interests." Economic power, in other words, is the preferred instrument of foreign policy in the 1990s and beyond.

Ironically, SOA defender Joseph C. Leuer may in fact strengthen the case for closing the School of the Americas when he stresses the centrality of economic power. Leuer describes how the SOA served "long-term foreign policy objectives" in Latin America as circumstances changed over many decades. During what he calls the "competitive bipolar world" of rivalry between the United States and the Soviet Union, the School of the Americas and U.S. foreign policy offered "support for many oligarchic Latin American regimes," including "authoritarian regimes" guilty of "human rights abuses." In "today's unipolar world," however, "the preponderance of armed conflicts decreased and economic, not ideological, factors came to the forefront of U.S. foreign policy."[118] The Bush administration, according to Leuer, exercised leverage through the "Enterprise for the Americas" whose "main effort was to change the perception of the Central American leaders as to the advantages and disadvantages of protectionism and widespread direct government intervention in the economy versus systems based on free competitive markets and the private sector." In order "to capitalize on the successes of the Enterprise for the Americas," he continues, "Bush and later Clinton formulated . . . [and] Congress ratified the North American Free Trade Agreement (NAFTA) which has to date created steady regional growth."[119]

Economic power became the most prominent element within low-intensity conflict and within U.S. foreign policy generally for a variety of reasons. The Third World — in debt, dependent, and with nowhere to turn — was particularly vulnerable to economic pressure. Nicaragua, which had tried to find a way between oppressive capitalism and state-dominated socialism, had, by the end of the bloody 1980s, been successfully destroyed. A decade or more of structural adjustment programs had forced many third world governments into economic straitjackets with far-reaching consequences. They were allowed to hold elections but their emerging democracies were robbed of content, their treasuries drained, and their economies placed in service to multinational companies and banks. Finally, the symbolic tearing down of the Berlin Wall and the actual collapse of the Soviet Union meant, among other things, that the "new world order" would be dominated by the global economy and those with the power to shape it. For this reason, President Clinton referred to the North American Free Trade Agreement (NAFTA) as the principal foreign policy issue of his administration.

It is in this context of a "unipolar world" in which democracy is subservient to and undermined by economic power that Leuer states:

> Current international political situations have allowed a critical look at U.S. foreign policy in Latin America when bipolar blinders were worn. Without the threat which created such thinking, it is now possible to view a "New World Order" as it pertains to Latin America and U.S. involvement. The developing policies and their focus on creating external and internal markets are reshaping U.S. and Latin American roles in a global economic atmosphere. Where once Latin

> American allies were judged by their anticommunist fervor, they will now be judged by their contribution to regional harmony and democratic progress, to include human rights.[120]

The recent flurry of editorial honesty concerning the School of the Americas is directly linked to the emergence of economic leverage as the preferred means of power projection for U.S. foreign policy. The United States has many choices concerning the conduct of foreign policy which make it possible for the School of the Americas to make cosmetic changes in its curriculum. This is why human rights and democracy have suddenly found a place within the school's smorgasbord of course offerings. Unfortunately for SOA supporters, economic leverage provides U.S. policy makers with so many choices that the school itself may be expendable. It is for this reason that many voices in the U.S. foreign policy establishment and the mainstream media express doubts about the *present* usefulness of the School of the Americas. These doubts are cast in light of abhorrence over the *past* conduct of SOA graduates and, at times, the SOA itself. Unfortunately, however, the debates over *present utility* and laments over *past abuses* avoid what is most needed: *an honest assessment of the often deadly impact of U.S. foreign policy past and present.* In other words, *closing the School of the Americas, while important, could itself be part of a strategy of damage control.*

As we bring this chapter to a close it is important that we recognize that closing the School of the Americas will be a hard-fought and important symbolic victory. The campaign to close the school, however, must be part of an honest assessment of U.S. foreign policy, past and present. The compelling reasons for changing U.S. foreign policy, and not merely closing the School of the Americas, are legion:

• The School of the Americas is only one of many places where soldiers from Latin America and the Caribbean are trained. General Raoul Cédras, leader of the military junta that ruled Haiti from 1991 to 1994, and his police chief, Major Joseph-Michel François, received training at Fort Benning, although the School of the Americas denies they are SOA graduates. Each played important roles in the overthrow of the democratically elected Haitian president and popular priest, Jean-Bertrand Aristide. Various U.S. government investigations into the murder of the Jesuit priests raise the issue of U.S. training of murderers, not only at Fort Benning but elsewhere, including at Fort Bragg.[121] Other SOA graduates are CIA operatives and the CIA itself is implicated in numerous atrocities. President Clinton not only supports the School of the Americas, he has significantly increased the budget of the Central Intelligence Agency which now exceeds $30 billion annually.

SOA critic and former instructor U.S. Army Major Joseph Blair says that any "worthwhile professional military course required by Latin American countries is readily available at our military service schools and training centers."[122] The opposite, unfortunately, is equally true. The School of the Americas could be closed and its deadly functions carried out elsewhere unless closing the school is linked to a fundamental reassessment of U.S. foreign policy, including a dramatically diminished and changed role for the Central Intelligence Agency.

• U.S. foreign policy, including School of the Americas support for terror and torture, has damaging domestic implications. Two examples can illustrate how U.S. citizens suffer the consequences of bad foreign policy. The first concerns drug trafficking. Senator John Kerry's investigation, "Drugs, Law, and Foreign Policy," concluded:

> Foreign policy considerations have interfered with the United States' ability to fight the war on drugs. Foreign policy priorities . . . halted or interfered with U.S. law enforcement efforts to keep narcotics out of the United States. Within the United States, drug traffickers have manipulated the U.S. judicial system by providing services in support of U.S. foreign policy. U.S. officials involved in Central America failed to address the drug issue for fear of jeopardizing the war effort against Nicaragua.[123]

The *San Jose Mercury News* investigated the CIA's role in allowing crack cocaine to be introduced into ghettos throughout the United States. The CIA, according to the reports, collaborated with a Nicaraguan drug dealer and subverted a Drug Enforcement Administration investigation because the drug-kingpin in question was using funds from crack sales to support the Nicaraguan contras:

> Danilo Blandon, a year-long *Mercury News* investigation found, is the Johnny Appleseed of crack in California. . . . On a tape made by the Drug Enforcement Administration in July 1990, Blandon casually explained the flood of cocaine that coursed through the streets of South-Central Los Angeles during the previous decade. . . . But unlike the thousands of young blacks now serving long federal prison sentences for selling mere handfuls of the drug, Blandon is a free man today. He has a spacious new home in Nicaragua and a business exporting precious woods, courtesy of the U.S. government, which has paid him more than $166,000 over the past 18 months. . . . In

> recent court testimony, Blandon, who began dealing cocaine in South-Central L.A. in 1982, swore that the first kilo of cocaine he sold in California was to raise money for the CIA's army, which was trying... to unseat Nicaragua's new socialist Sandinista government.[124]

Another linkage between U.S. foreign policy, the CIA, drugs, and the contras leads to Oliver North and to School of the Americas graduate Juan Rafael Bustillo. According to former Drug Enforcement Administration agent Celerino Castillo III, Oliver North was the "leader... of Latin America's most protected drug smuggling operation."[125]

> Contra planes flew north to the U.S., loaded with cocaine, then returned laden with cash. All under the protective umbrella of the United States Government. My informants were perfectly placed: one worked with the Contra pilots at their base, while another moved easily among the Salvadoran military officials who protected the resupply operation. They fed me the names of Contra pilots. Again and again, those names showed up in the DEA database as documented drug traffickers. When I pursued the case, my superiors quietly and firmly advised me to move on to other investigations.[126]

At the behest of U.S. leaders, SOA graduate General Bustillo, who headed the Salvadoran Air Force, gave the contras "carte blanche" to ship weapons to Nicaragua and narcotics to the United States. CIA operative and Bay of Pigs veteran Felix Rodríguez "headed the operation and reported to Oliver North."[127] When General Bustillo was passed over for a cabinet post in El Salvador because of U.S. opposition, DEA agent Castillo reports that Bustillo "flew into a rage, asking how the U.S. government could stab him in the back after what he did for the Contras."[128]

The second issue which illustrates how U.S. citizens suffer the consequences of bad foreign policy concerns terrorism. U.S. foreign policy and the School of the Americas, as we have seen, have been closely tied to practitioners and policies of torture and terror. Terrorism may come home to roost. The Oklahoma City bombing was a horrible event. As it turned out, the perpetrators of the ghastly deed were members of a white militia group. Julio Noboa, in a commentary in the *San Antonio Express News* entitled, "Let's stop training terrorists," writes in the context of the Oklahoma City bombing:

> The fact that this vengeful violence against innocents happened right here at home surprised and shocked many of us. Yet the terrorism generated over the last few decades against our neighbors to the south has been far more destructive. Hundreds of thousands of Latin American families have been subjected to kidnappings, torture and murder. . . . What should be shocking to us all is that our tax dollars have been spent in training some of these terrorists right here at home, in the Army base at Fort Benning, Ga.[129]

Noboa doesn't say it, but the ideology which grounded the actions of those responsible for the Oklahoma City bombing could easily have been learned at the School of the Americas. And isn't it possible that terrorist acts within the United States could one day be carried out by people from the Americas for whom terrorism might be seen as a legitimate response to the policies and practices of U.S. foreign policy and the School of the Americas?

• Finally, it should give us pause that poverty and social inequality, key causes of the social turmoil throughout

Latin America and much of the Third World since the end of World War II, are deepening as a result of past policies and present structural adjustments. Jesuit economist Xabier Gorostiaga cites United Nations statistics indicating that the net worth of 358 billionaires is greater than the yearly incomes of 45 percent of humanity and that 60 percent of the world's people are living and dying on 6 percent of the world's wealth.[130] The gap separating the rich from the poor, both within and between nations, is growing rapidly. Injustice and stability are still irreconcilable. When economic leverage no longer keeps people and nations in line the United States will likely revert to cruder methods of terror and torture. Perhaps this is one of the reasons the School of the Americas still has its defenders.

After the fall of the Berlin Wall, General A. M. Gray, commandant of the Marine Corps, argued against significant cuts in U.S. military spending and on behalf of an ongoing role for the Marine Corps in the post-Cold War world. His argument could be used by SOA supporters as well:

> The underdeveloped world's growing dissatisfaction over the gap between rich and poor nations will create a fertile breeding ground for insurgencies. These insurgencies have the potential to jeopardize regional stability and our access to vital economic and military resources. This situation will become critical as our nation and allies and potential adversaries become more and more dependent on these strategic resources. If we are to have stability in these regions, maintain access to their resources, protect our citizens abroad, defend our vital installations, and deter conflict, we must maintain within our active force structure a credible military power projection capabil-

> ity with the flexibility to respond to conflict across the spectrum of violence throughout the globe.[131]

When I visited El Salvador in the spring of 1996 I was told by a prominent religious leader that the United Nations as part of its Truth Commission report presented the United States and Salvadoran governments with the names and addresses of Salvador's death squads. The U.N. quietly demanded that the death squads be dismantled. According to my source, which I trust, they haven't been touched and the United States seems to be holding them in reserve in case things begin to get out of hand once again in El Salvador.

Damage control won't work anymore. The cat is out of the bag. Closing the School of the Americas is a realistic goal for those of us troubled by the conduct of the school and its graduates. But closing the School of the Americas, though vitally important, is not enough. We must close the School of the Americas as we work for fundamental shifts in U.S. foreign policy.

Chapter 8

Close the School Now!

> I'm a United States citizen, I pay federal taxes. My taxes are complicit in funding the School of the Americas which teaches Latin America military techniques of anti-insurgency, which translates into murders, disappearance, torture, atrocities. I have a measure of complicity in what that school does in my name with my tax money.
>
> —SOA DEFENDANT EDWARD F. KINANE[132]

I thought of our nation's historical amnesia when I read through the transcripts of the trial of thirteen SOA Watch supporters. On April 29, 1996, thirteen defendants went before District Court Judge Robert Elliott. They were arrested on November 16, 1995, for an action at the School of the Americas in which they reenacted the murder of the six Jesuit priests, their housekeeper and her daughter. Their action both commemorated the grisly murders which had occurred on that same day, six years prior, and called attention to the school's role in training those responsible for this and numerous other massacres and human rights violations.

What we refuse to know can be hurtful to ourselves and to others. If it is true that the truth can set us free, then is it not also true that lies, secrecy, deception, and self-serving distortions can undermine our freedom? During the course of the trial of SOA protesters the following exchange took place involving a witness, defense and prosecution lawyers, and the judge:

> *Defense Attorney Peter Thompson:* Was part of your purpose on November 16, '95 to call attention to those concerns about torture being committed by those trained at the School of the Americas?
>
> *Defendant Edward F. Kinane:* Was it part of my concern? Definitely.
>
> *Defense Attorney Peter Thompson:* Was it part of your message?
>
> *Defendant Edward F. Kinane:* Definitely.
>
> *Defense Attorney Peter Thompson:* And have you recently observed something that has confirmed and [is] related to those concerns about torture?
>
> *Defendant Edward F. Kinane:* Yes. I've seen a video of an interview with a graduate of the School of the Americas in which he acknowledges that during his training he was taught the skills of torture.
>
> *Prosecuting Attorney George F. Peterman:* Your Honor, I'm going to object. This is clearly outside the testimony.
>
> *Judge Robert Elliott:* Yes. *Let's skip over that.*
>
> *Defense Attorney Peter Thompson:* For the record, I'm showing you Defendants' Exhibit 10. Is this the video that you saw?
>
> *Defendant Edward F. Kinane:* . . . yes.
>
> *Prosecuting Attorney George F. Peterman:* We object, Your Honor.
>
> *Judge Robert Elliott:* I sustain the objection. (Emphasis added.)[133]

The prosecuting attorney's objections and the judge's responses are powerful symbols of our refusal to know, our stubborn will to keep hidden what must be revealed if we are to understand the deadly legacy of the School of the Americas and its relationship to broader U.S. foreign policy objectives. "Let's skip over that" seems a perfect summary of our nation's historical amnesia. It cloaks the actions of the School of the Americas in secrecy and contributes to an attitude of disbelief when groups like SOA Watch succeed in bringing ugly revelations to light.

"Let's skip over that." Let's hide beneath the cloak of secrecy as long as possible and when that is no longer tenable let's retreat into comfort zones lined with explanations based on a "few bad apples" or "rogue elements" or other theories which gloss over whatever reprehensible conduct we find objectionable and confine it to the realm of the exceptional. The conduct of the School of the Americas and its graduates is somehow divorced from U.S. foreign policy, our basic values, or our intent.

The thirteen defendants who went before Judge Elliott that day could no longer "skip over" the blood-soaked evidence linking the School of the Americas to terror and torture. There was JoAnne Lingle, a fifty-nine-year-old mother who, while in Guatemala, was "moved by the mothers' testimonies . . . mothers who had lost their children through the violence . . . mothers who had found their mutilated bodies by the roadside or often in garbage dumps." Lingle also could not get another mother, Rufina Amaya, out of her mind. Amaya, the sole survivor of the El Mozote massacre in El Salvador, had witnessed the slaughter of more than nine hundred people, including four of her children who were bayonetted to death at the hands of SOA graduates. Amaya had heard her nine-year-old son call out to her, "Mama, they are killing us." JoAnne Lingle heard this cry too and could not be silent. Explaining her efforts to close the School of the Americas, she said, "I want to stop the suffering. I want to stop the violence."

Another SOA defendant was Edward Kinane. Concerned about his tax dollars and more, Kinane had spent five months as the unarmed personal bodyguard to Medardo Goméz. Goméz, the Lutheran bishop in El Salvador, had received numerous death threats. "I have seen terror in the eyes of peasants in El Salvador during the war," Kinane said. "They were frightened of the Salvadoran military — a military in which, in large part, the officers had been trained at the School of the Americas."

There was also Sister Claire O'Mara, a seventy-four-year-old Ursuline nun, who had worked in Peru, Mexico, and the South Bronx. She had a large poster of the four U.S. churchwomen murdered in El Salvador hanging on her wall for almost sixteen years. She could not forget them. "Their deaths took place on December 2, 1980 — just ten months after Archbishop Oscar Romero had also been murdered in El Salvador," she reminded the court. "It seemed to me," she said, explaining why she engaged in civil disobedience at Fort Benning in an effort to close the School of the Americas, "a very small contribution to the cause of justice that I should risk something as unimportant as my own person in order to be a witness to the fact that there is a higher law and that human dignity — whatever the country — must be respected."[134]

"Let's skip over that." The judge had the power to impose historical amnesia in the courtroom. But we have the power to say no. We can't skip over that. We won't. Too much blood has been shed. Too much senseless death tolerated. Too many lies propagated. Too much is at stake. What I have done in previous chapters, borrowing language from the court, is to put the School of the Americas on trial. The evidence presented against the school, it seems to me, makes an airtight case with the possible exception of an insanity plea.

Let's review the case against the School of the Americas. The school is on trial for the murder and torture of thousands of citizens throughout Latin America since its founding in 1946. It is accused of aiding and abetting assassins, dictators, and thugs and of a massive coverup.

In the course of our "trial" we have heard defense witnesses attempt to refute these charges, utilizing one or more of the following arguments: The School of the Americas should not be condemned due to the inappropriate actions of "a few bad apples"; the curriculum, especially recent changes, demonstrates that the central goals of the school are to promote human rights, to encourage democracy, and to professionalize Latin American armies; exposing Latin American soldiers to American values and culture at the school reinforces democracy; the SOA provides a common table where U.S. military leaders and their Latin American counterparts can share perspectives and interests; and finally, the School of the Americas serves vital U.S. interests.

All of these defense arguments, with the exception of the last one concerning vital interests, were rebutted. Witnesses for the prosecution demonstrated that the record of abuse was so extensive and far-reaching that it made the "bad apple" argument untenable. SOA graduates can be traced to nearly every coup and human rights travesty to occur in Latin America over the past five decades. We placed so many SOA graduates at the scene of grisly crimes as to establish an irrefutable link between the school and torture and other human rights abuses. On many occasions SOA graduates were literally holding U.S. weapons, their hands dripping with blood.

The link between the School of the Americas, SOA graduates, and terror firmly established, we then sought to establish motive. In the course of our "trial" we allowed many defense witnesses to incriminate themselves. In their testimony, using their own words, we heard about "implacable enemies" and about the need "to subvert, sabotage, and destroy our enemies by more clever, sophisticated, more effective methods than those used against us." We heard them equate liberation theology with subversion. Their rhetoric was both revealing and ugly. And the policies which grew out of their venomous words included assassination and torture.

We described numerous crimes and placed SOA graduates at all of them. We demonstrated both opportunity and motive. As stated previously, SOA graduates used U.S. weapons when committing most of their atrocities. If this "smoking gun" wasn't sufficient, we brought forth SOA training manuals which advocate executions, torture, false arrest, blackmail, censorship, payment of bounty for murderers, and other forms of physical abuse. And finally, former SOA graduates testified that they were taught torture techniques at the School of the Americas.

The only point of agreement between witnesses defending the School of the Americas and those advocating closure is that the school "is an implement of foreign policy" which served vital U.S. interests. The terror, the torture, the targeting of progressive religious, the human rights abuses, the senseless deaths were all linked to official policies. At the beginning of this book I suggested that the School of the Americas had done its job either very poorly or very well. In either case it needs to be closed. The evidence suggests that the school has done its job efficiently while carrying out its horrific mission. Training Latin American soldiers in the craft of counterinsurgency, including the arts of terror and torture, was deemed vital and necessary. Some say the school is no longer necessary but that misses the point. If the School of the Americas is linked to terror and torture, and if terror and torture were utilized in service to policies defending vital interests, which I think is undeniable, then it is U.S. foreign policy that is ultimately on trial.

The School of the Americas must be closed and momentum to close it is building. We have already seen a wave of editorials in newspapers big and small condemning the School of the Americas. The Leadership Conference of Women Religious and the Presbyterian Church U.S.A. have passed resolutions calling for SOA closure. In August 1996 four hundred Catholic sisters held a prayerful protest, adding their voices to the thousands of others calling for a shutdown of the SOA. I attended a demonstration of more than four hundred people at SOA headquarters in November 1996. Sixty-two people were arrested in a symbolic protest. A year earlier there were twenty-three protesters and thirteen arrests. Next year there will be many more people and arrests if the school remains open. Several legislative efforts to close the school have been tried and failed. They will be tried again. Eventually they will succeed.

You can help. Write. Call. Organize. Show the powerful video from SOA Watch depicting the many crimes of the School of the Americas. Encourage your local paper to take an editorial stand. Get organizations to which you are connected to speak out. Arrange for speakers from SOA Watch to visit your community. Your voice matters. Together we can succeed. It will not be easy but the School of the Americas can and will be shut down.

The day after the School of the Americas is shut down we can have a wonderful party. Let's celebrate. Let's thank Father Roy Bourgeois, Diana Ortiz, and the multitude of faithful witnesses for not letting our nation or ourselves "skip over that." Let's write to the newspapers that took a stand and thank them for their honesty. Let's tell Representative Joseph Kennedy that we appreciate his persistence. Let's even congratulate ourselves for joining with others, for taking a stand, for speaking out. The day after our celebration, however, let's return to work. As I said previously, it is U.S. foreign policy that is on trial. Let's work to end the CIA as we know it. Let's work for domestic and foreign policies which place at their center the goal of ending poverty. Let's create stability by first establishing justice. If there is one lesson we must learn from all the carnage traceable to U.S. foreign policy and to the School of the Americas in the post-World War II period it is this: stability without a decent measure of justice is an impossible goal. Its pursuit necessarily leads to a spiral of escalating violence. We would do well to remember that lesson when we consider that sometime around the year 2050, if globalization proceeds on its present course and if U.S. inequality continues to deepen, nearly half of the world's people will live in absolute poverty and half of the U.S. people will be in jail.[135]

The final word appropriately belongs to Father Roy Bourgeois. His faithful, prophetic, persistent witness has filled the campaign to close the School of the Americas with integrity and hope. As Judge Elliott was about to sentence him to six more months in federal prison, Father Bourgeois told the court:

> Prison is hard. I have been there and others have been there. But let me just say: If going to prison will help close the doors of the School of the Americas, we go. We go.
>
> We go in solidarity with others, with people like Archbishop Romero. We go in solidarity with the four churchwomen whom Sister Claire and others have

talked about. We go in solidarity with sisters and brothers whom we haven't read about in our newspapers. They are the nameless, the unknowns. But they have been victims of the School of the Americas.

We want this School closed because it is a School of thugs. It is a School of terrorists. It is a School that brings shame upon our country and upon us and our laws and what we should be standing for.

I'm filled with hope, really. We came here to Columbus, our very small group, six years ago. At our first meeting we had three people. It was very lonely in those days. It still gets lonely in the struggle for peace and justice. But the word about this School of the Americas is traveling around our country and it's uniting people of all different ages, of all different walks of life. People are coming together hearing about this School in Columbus, Georgia, that is bringing shame upon Fort Benning and our armed forces and our country.

People will continue to come here in greater numbers each year to say: "No, not in our name will we allow you to keep this school going." I have no doubt at all that one day, and I pray that it's soon, this school, the School of the Americas which has caused so much suffering and death to our sisters and brothers abroad and has been a theft from the poor here at home, this school will close. We will not stop speaking out until it does.

We will speak from prison, Your Honor. We will speak from our cells. The truth cannot be silenced, it can't be chained.[136]

Notes

1. The Center for Global Education is a program of Augsburg College in Minneapolis, Minnesota. The Center offers semester programs for college students in Mexico, Central America, and Namibia. It also offers travel seminars, usually two weeks in length, to countries throughout the world. My wife and I co-directed the Center's House of Studies in Nicaragua from 1984 to 1986. The Center's experiential learning programs are among the best offered by any group in the United States.

2. Arthur Jones, "Haiti, Salvador Links Viewed," *National Catholic Reporter,* November 19, 1993.

3. Editorial by Frank del Olmo, *Los Angeles Times,* April 3, 1995.

4. For more information on SOA Watch and how you can help in the campaign to close the School of the Americas write to SOA Watch, P.O. Box 3330, Columbus, GA 31903.

5. Quoted from the "Transcript of Trial," United States District Court, Middle District of Georgia, Columbus Division, in the case of "The United States of America *v.* William J. Bichsel, et al.," April 29, 1996, p. 199.

6. *Congressional Record,* May 20, 1994, p. H3771.

7. "The U.S. Army School of the Americas," a brochure available from Commandant, U.S. Army School of the Americas, ATTN: PAO, Fort Benning, GA 31905.

8. Ibid.

9. Ibid.

10. "School of the Americas and U.S. Foreign Policy Attainment in Latin America," an "information paper" by Joseph C. Leuer, January 1996, p. 1.

11. Ibid., pp. 8–10.

12. Ibid., pp. 11–14.

13. For more information contact SOA Watch, P.O. Box 3330, Columbus, GA 31903.

14. *United Nations Truth Commission Report,* March 15, 1993.

15. Letter from Representative Martin Meehan to Defense Secretary Les Aspin, August 6, 1993.

16. "Our Man in Guatemala," *Washington Post,* March 26, 1995.

17. "The Vigil Begins," excerpted from *Sojourners,* July–August 1996, p. 18.

18. Ibid., pp. 18 and 19.

19. Ibid., p. 18.

20. Ibid., p. 19.

21. Ibid.

22. From an editorial, "The Chickens Come Home to Roost," *In These Times,* April 15, 1996, p. 2.

23. Daniel Maloney, "SOA Recognizes 1991 Staff College Graduates," *The Bayonet,* Friday, January 3, 1992.

24. "Our Man in Guatemala," *Washington Post,* March 26, 1995.

25. Tim Weiner, "A Guatemala Officer and the C.I.A.," *New York Times,* March 26, 1995.

26. *Congressional Record,* May 20, 1994, p. H3771.

27. Douglas Waller, "Running a 'School for Dictators,'" *Newsweek,* August 9, 1993, pp. 34 and 37.

28. Ibid.

29. "A School for Closing," *Cleveland Plain Dealer,* July 20, 1995.

30. "Running a 'School for Dictators,'" *Newsweek,* August 9, 1993.

31. *Washington Post,* May 23, 1994.

32. *Columbus Ledger-Enquirer,* March 29, 1994.

33. "School of the Americas and U.S. Foreign Policy Attainment in Latin America," an "information paper" by Joseph C. Leuer, January 1996, p. 12.

34. Ibid., pp. 12–13.

35. "School of Assassins," Roy Bourgeois, MM, in *Blueprint for Social Justice* 48, no. 8 (April 1995), p. 2. *Blueprint* is a newsletter published by the Twomey Center for Peace through Justice at Loyola University.

36. *House Congressional Record,* May 20, 1994, p. H3771.

37. Charles T. Call, "Academy of Torture," *Miami Herald,* August 9, 1993.

38. "Running a 'School for Dictators,'" *Newsweek,* August 9, 1993, p. 37.

39. Tim McCarthy, "School Aims at Military Control," *National Catholic Reporter,* April 8, 1994.

40. Ibid.

41. Joseph Blair, "SOA Isn't Teaching Democracy," *Columbus Ledger-Enquirer,* July 20, 1993.

42. "School Aims at Military Control."

43. "Running a 'School for Dictators,'" *Newsweek,* August 9, 1993, p. 37.

44. "School of the Americas and U.S. Foreign Policy Attainment in Latin America," p. 13.

45. "School Aims at Military Control," *National Catholic Reporter,* April 8, 1994.

46. *House Congressional Record,* May 20, 1994, p. H3771.

47. *Inside the School of Assassins,* produced by Robert Richter, Richter Productions. This hour-long documentary can be ordered from SOA Watch.

48. Gary Cohn and Ginger Thompson, "Unearthed: Fatal Secrets," *Baltimore Sun,* reprint of a series that appeared June 11–18, 1995.

49. Ibid.

50. *Inside the School of Assassins,* video.

51. "Lessons in Terror," *Boston Globe,* October 1, 1996.

52. "School Aims at Military Control."

53. Jon Sobrino, *Companions of Jesus: The Jesuit Martyrs of El Salvador* (Maryknoll, N.Y.: Orbis Books, 1990), p. xviii.

54. "Shut It Down," *Syracuse Post-Standard,* May 2, 1994.

55. "The U.S. Army School of the Americas," a brochure

available from Commandant, U.S. Army School of the Americas, ATTN: PAO, Fort Benning, GA 31905.

56. This quotation is taken from the written transcript of a Public Affairs Television special, with Bill Moyers, entitled *The Secret Government: The Constitution in Crisis.* The program was a production of Alvin H. Perlmutter, Inc., and Public Affairs Television, Inc., in association with WNET and WETA. Copyright 1987 by Alvin H. Perlmutter, Inc., Public Affairs Television, Inc. The written transcript was produced by Journal Graphics, Inc., New York.

57. Speech before the National Foreign Trade Convention, November 12, 1946.

58. Quoted in Michael T. Klare and Peter Kornbluh, eds., *Low Intensity Warfare: Counterinsurgency, Proinsurgency, and Antiterrorism in the Eighties* (New York: Pantheon Books, 1988), p. 48.

59. "School of the Americas and U.S. Foreign Policy Attainment in Latin America," p. 7.

60. Ibid., p. 1.

61. Ibid.

62. Ibid., p. 2.

63. Ibid., pp. 2–3.

64. *Atlanta Constitution,* June 3, 1995.

65. Jonathan Marshall, Peter Dale Scott, and Jane Hunter, *The Iran Contra Connection: Secret Teams and Covert Operations in the Reagan Era* (Boston: South End Press, 1987), p. 31.

66. José Comblin, *The Church and the National Security State* (Maryknoll, N.Y.: Orbis Books, 1979), p. 65.

67. "U.S., Latin America Sign Secret Defense Plan," *National Catholic Reporter,* December 16, 1988.

68. Jack Nelson-Pallmeyer, *Brave New World Order: Must We Pledge Allegiance?* (Maryknoll, N.Y.: Orbis Books), 1992, pp. 35–40.

69. Eric Black, *Rethinking the Cold War* (Minneapolis: Paradigm Press, 1988), pp. 9–10.

70. Ibid., p. 10.

71. Ibid., pp. 10–11.

72. Tom Tomorrow, "This Modern World," *Des Moines Sunday Register,* May 14, 1995.

73. "Report on the Guatemala Review," Intelligence Oversight Board, June 28, 1996, p. 84.

74. Lisa Haugaard, "Torture 101," *In These Times,* October 14, 1996, pp. 14–16.

75. Dana Priest, "U.S. Instructed Latins on Executions, Torture," *Washington Post,* September 21, 1996.

76. "Torture 101," *In These Times,* p. 14.

77. Joseph Blair and I spoke at this demonstration to close the School of the Americas. This quote is from his talk to protesters on November 16, 1996.

78. Jack Nelson-Pallmeyer, *War against the Poor: Low-Intensity Conflict and Christian Faith* (Maryknoll, N.Y.: Orbis Books, 1989), p. 15.

79. Tim McCarthy, "School Aims at Military Control," *National Catholic Reporter,* April 8, 1994.

80. America's Watch testimony, January 31, 1990.

81. Quoted in Suzanne Gowan et al., *Moving Toward a New Society* (Philadelphia: New Society Press, 1976), pp. 86–87.

82. *Atlanta Constitution,* June 3, 1995.

83. Penny Lernoux, *Cry of the People* (New York: Penguin Books, 1980), pp. 142–43.

84. The Committee of Santa Fe, "A New Inter-American Policy for the Eighties" (Washington, D.C.: Council for Inter-American Security, 1980).

85. This excerpt from the U.S. military training manual is found in Lisa Haugaard, "Torture 101," *In These Times,* October 14, 1996, p. 16.

86. Ibid, p. ii.

87. Ibid.

88. Tim McCarthy, "School Aims at Military Control," *National Catholic Reporter,* April 8, 1994.

89. Quoted in *Total War against the Poor* (New York: New York Circus Publications, 1990), p. 133.

90. Bob Woodward, *Veil: The Secret Wars of the CIA* (New York: Simon & Schuster, 1987), p. 195.

91. Quoted in an article by Michael Klare, "Low Intensity Conflict: The War of the 'Haves' against the 'Have Nots,'" *Christianity & Crisis,* February 1, 1988, pp. 12–13.

92. Ibid.

93. Woodward, *Veil: The Secret Wars of the CIA,* p. 173.

94. "Contra Forces Target Civilian Medical Work in Northern Nicaragua," Executive Summary Report, U.S. Medical Task Force Investigation, January 1988.

95. Ibid.

96. This quotation is taken from the written transcript of a Public Affairs Television special, with Bill Moyers, entitled "The Secret Government: The Constitution in Crisis." The program was a production of Alvin H. Perlmutter, Inc., and Public Affairs Television, Inc., in association with WNET and WETA. Copyright 1987 by Alvin H. Perlmutter, Inc., Public Affairs Television, Inc. The written transcript was produced by Journal Graphics, Inc., New York, p. 14.

97. Jack Nelson-Pallmeyer, *War against the Poor: Low-Intensity Conflict and Christian Faith* (Maryknoll, N.Y.: Orbis Books, 1989), p. 33.

98. "Affidavit of Edgar Chamorro," Case concerning Military and Paramilitary Activities in and against Nicaragua (Nicaragua *v.* United States of America), International Court of Justice, September 5, 1985, p. 21.

99. Quoted in William I. Robinson and Kent Norsworthy, *David and Goliath: The U.S. War against Nicaragua* (New York: Monthly Review Press, 1987), pp. 56–57.

100. See Jack Nelson-Pallmeyer, *War against the Poor.*

101. Martin Lange and Reinhold Iblacker, eds., *Witnesses of Hope: The Persecution of Christians in Latin America* (Maryknoll, N.Y.: Orbis Books, 1981), p. 74.

102. Celerino Castillo III and Dave Harmon, *Powderburns: Cocaine, Contras and the Drug War* (Oakville, Buffalo, London: Mosaic Press, 1994), pp. 151–54.

103. Ibid., p. 154.

104. Ibid., pp. 79–80.

105. Molly Ivins, "School of the Americas Training of Latin American Thugs Should Be Old News," *Star Tribune,* October 6, 1996.

106. *San Antonio Express-News,* April 14, 1995.

107. *Des Moines Register,* May 16, 1995.

108. *Atlanta Constitution,* June 3, 1995.

109. *New York Times,* March 24, 1995.

110. *Cleveland Plain Dealer,* July 20, 1995.

111. "Report on the Guatemala Review," Intelligence Oversight Board, June 28, 1996, p. 84.

112. Dana Priest, "U.S. Instructed Latins on Executions, Torture," *Washington Post,* September 21, 1996.

113. "School of the Dictators," *New York Times,* September 28, 1996.

114. Ibid.

115. Theotis Robinson, "Taxpayers Fund 'School for Assassins,'" *Knoxville News-Sentinel,* August 7, 1995.

116. "Close This School," *Bangor Daily News,* October 3, 1995.

117. "Our Man in Guatemala," *Washington Post,* March 26, 1995.

118. "School of the Americas and U.S. Foreign Policy Attainment in Latin America," an "information paper," pp. 1 and 4.

119. Ibid., pp. 4–5.

120. Ibid., p. 5.

121. See, for example, the "Special Task Force on El Salvador" headed by Joseph Moakley which submitted its report on April 10, 1990.

122. "SOA Isn't Teaching Democracy," *Columbus Ledger-Enquirer,* July 20, 1993.

123. "Drugs, Law, and Foreign Policy," a report from the Senate Foreign Relations Committee's Subcommittee on Narcotics, Terrorism, and International Operations, April 1989.

124. Gary Webb, "War on Drugs' Unequal Impact on U.S. Blacks . . . ," *San Jose Mercury News,* August 20, 1996.

125. Celerino Castillo III and Dave Harmon, *Powderburns: Cocaine, Contras and the Drug War* (Oakville, Buffalo, London: Mosaic Press, 1994), p. 128.

126. Ibid., p. 22.

127. Ibid., p. 128.

128. Ibid., p. 210.

129. *San Antonio Express News,* May 20, 1995.

130. Xabier Gorostiaga, "World Has Become a 'Champagne Glass,'" *National Catholic Reporter,* January 27, 1995.

131. A. M. Gray, "On the Corps' Continuing Role," *Marine Corps Gazette,* May 1990, p. 19.

132. Quoted from the "Transcript of Trial," United States District Court, Middle District of Georgia, Columbus Division, in the case of "The United States of America *v.* William J. Bichsel, et al.," April 29, 1996, p. 109.

133. Ibid., pp. 109–10.

134. Excerpted from "The Truth Cannot Be Silenced," the trial testimonies of the SOA 13. Available from SOA Watch, P.O. Box 3330, Columbus, GA 31903.

135. For more details see my recent book, *Families Valued: Parenting and Politics for the Good of All Children* (New York: Friendship Press, 1996).

136. "The Truth Cannot Be Silenced," the trial testimonies of the SOA 13.